"This Road Doesn't Go to My Apartment!"

"I know. We're not going to your apartment. You're coming home with me."

"I suppose you've decided that I should reward your generosity?" Nikki responded acidly.

"Nikki," Michael said in an icy voice, "I enjoy women who are *anxious* to please me. Nonetheless, next time you throw yourself at me, I'm going to accept the invitation."

Later, looking around "her" room, a treacherous thought entered her head. How easy it would be to stop fighting, to agree to whatever Michael wanted of her. But for how long? And how would it feel to fall in love with him, only to be traded in like last year's model?

BROOKE HASTINGS
is an avid reader who loves to travel. She draws her material from many sources: the newspaper, politics, the places she visits and the people she meets. Her unique plots, full of real people who meet love in many guises, make her one of the best new writers in this field.

Dear Reader:

Silhouette Romances is an exciting new publishing venture. We will be presenting the very finest writers of contemporary romantic fiction as well as outstanding new talent in this field. It is our hope that our stories, our heroes and our heroines will give you, the reader, all you want from romantic fiction.

Also, *you* play an important part in our future plans for Silhouette Romances. We welcome any suggestions or comments on our books and I invite you to write to us at the address below.

So, enjoy this book and all the wonderful romances from Silhouette. They're for *you*!

Karen Solem
Editor-in-Chief
Silhouette Books
P.O. Box 769
New York, N.Y. 10019

BROOKE HASTINGS
Playing for Keeps

Silhouette *Romance*

Published by Silhouette Books New York

Distributed in Canada by PaperJacks Ltd., a Licensee
of the trademarks of Simon & Schuster, a division of
Gulf+Western Corporation.

SILHOUETTE BOOKS, a Simon & Schuster Division of
GULF & WESTERN CORPORATION
1230 Avenue of the Americas, New York, N.Y 10020

Copyright © 1980 by Brooke Hastings

Distributed by Pocket Books

ISBN: 0-671-57013-7

First Silhouette printing June, 1980

10 9 8 7 6 5 4 3 2 1

Printed in Canada

Playing for
Keeps

Chapter 1

"Damn!" Nikki Warren muttered groggily. For a moment she stared balefully at the digital clock, which read 7:33. Finally, reluctantly, she dragged herself out of bed. Monday morning. It had been an exhausting weekend, running back and forth between her mother's hospital room and the neighborhood store her mother owned. But what drained Nikki's usually ample energy was not physical activity, but a pervasive sense of helplessness, powerlessness.

She was used to managing her life decisively and rationally, making her own luck. At the age of twenty-three, it was sobering for her to learn that Nicola Warren could not even dictate her personal future, much less anyone else's, not even to the extent of shutting off the worried flow of thoughts that had kept her from falling asleep last night and so many nights before. And not even to the extent of getting to work on time this morning.

This had been a bitterly cold winter in New York City. Although according to the calendar it would be spring in less than two weeks, crusty snow was still piled up along the side streets; icy spots lurked on the sidewalks. The most recent storm had swept down from Canada on Friday, dumping almost eight inches of new

snow on a battered city during the night. As protection
against the winter cold she hastily pulled on a beige
wool turtleneck and a wine-colored vest and tucked her
matching slacks into ugly but functional fur-lined black
boots. She threw a pair of low-heeled shoes into a
shopping bag to take with her. Long silky brown hair
was expertly wound into a tight chignon and a little
makeup applied to a face that needed none at all.

Nikki gulped a glass of orange juice, decided there
was no time to make herself lunch to take along, pulled
on a midi coat, ski hat and woolen mittens, and flew out
the door of the apartment. It was 7:50.

Nikki negotiated the slippery sidewalk in record
time, only to arrive at the bottom of the subway steps
just as an express train was pulling away. She impatiently stood in the cold station and cursed as a train
that was supposed to stop flew past the platform
without so much as slowing down.

When the next subway finally arrived, it was a local
and it was packed. Sighing with resignation, Nikki
elbowed her way into the car. She was 5′ 3½″ in her
boots and everyone towered over her, making her feel
like a very small sardine in a very large can.

Only when she stood wedged among that mass of
bundled-up humanity on the subway did Nikki's mind
return to her problems. She stared at some of the other
passengers. Did any of them, Nikki mused, have
fantasies about robbing a bank? Probably not.

But then undoubtedly none of them had looked on
impotently as she had, and watched a mother's savings
dwindle to nothing, wiped out by illness. And now
hospital bills were mounting, and after that would
come the cost of therapy.

Mrs. Warren had been widowed six years before,
when Nikki was a high school senior of seventeen. The
Warrens owned a small store which carried a selection

of tasteful gifts and housewares; it was located in the upper Bronx, north of Manhattan's skyscrapers and excitement. When Nikki's father had died after a heart attack, Pamela Warren continued to work, running the store as she had with her husband. This modest business permitted a comfortable, if not luxurious, standard of living. She hired a part-time saleswoman when her daughter entered college the next fall.

Nikki had intended to enroll at a city school where tuition was low, but her mother had pressed her to use the life insurance money from her father's death for expenses at a prestigious private women's college to which Nikki had also been accepted. She had put up only token opposition because she was thrilled with the idea of being able to attend the school.

She commuted to Manhattan each day. She was a good student who very much wanted to make her mother proud of her. She confidently anticipated that a good academic record would earn her a well-paying job, so she put in long hours studying and in fact did exceptionally well.

After finishing college, Nikki had gotten a job in the advertising department of the New York *Sun*, an influential New York morning newspaper. The position was not an exceptionally challenging one, but she knew that the chances for promotion were excellent. And promotion was very much a part of her plan.

She found a place to live fairly close to her job, which was in mid-town Manhattan. The studio apartment consisted of an efficiency kitchen, bath and living room. She purchased a convertible couch to sleep on. Nikki preferred these cramped living quarters to sharing an apartment with a stranger or casual acquaintance.

Everything seemed perfect—especially when, after eight months, she applied for a position in the personnel department of the newspaper, and was accepted.

Nikki loved the new job, particularly the part of it that involved interviewing and testing applicants.

And then, a month ago, Pamela Warren had been felled by a cerebral vascular accident, the physicians' jargon for a stroke. She had no medical insurance and was too young to be eligible for government programs. For the first twenty-four hours as Mrs. Warren lay in a coma, a grief-stricken Nikki thought not at all about money. She was far too busy arranging for extra help in the store, sitting by her mother's bedside in the intensive care area, talking to the staff doctors. During the first week of her mother's hospitalization she spent hours riding the subway between hospital, store, work and her apartment. She was anesthetized by shock and weariness. To lose one beloved parent at seventeen had been painful enough, that she might now lose her mother did not seem credible.

When Pam Warren, true to her reputation as a fighter, began to recover more rapidly than anyone could have predicted or hoped, Nikki closed up her apartment and moved back to her mother's. It was closer to the store and hospital, if not to her job. Since then she felt as though her life had been lived on a treadmill.

As the train screeched to a halt at Forty-second Street, Nikki's natural optimism reasserted itself. She permitted herself to be swept out of the car along with the sea of exiting sardines. Nonchalantly dodging between the snarls of too many taxis, buses and cars, she told herself, all that matters is that mom is getting better. She could be out of the hospital in two weeks or even less. And I'll pay the bills somehow. It's all going to work out.

8:45. Nikki walked briskly from the elevator through the impersonal tiled corridor to her office on the fifth floor. She breathlessly opened the door and collided almost head-on with Erika Berger. Erika also worked in personnel placement and the two girls shared a small office. Erika had a snappy, New York sense of humor, and she and Nikki had become fast friends in the two months since they had met.

Erika cocked an eyebrow and looked down at her panting office-mate. "Well, you finally did something wrong. But don't panic. You won't be shot at dawn for coming in fifteen minutes late."

Nikki hurriedly whispered, "Did Mr. Martens notice?"

"Of course he did," Erika gibed. "He always notices where you are, or aren't, in this case. But he's concerned, not angry." She patted Nikki's shoulder. "Why don't you go in and hold his hand and tell him his favorite staff member is okay?"

Nikki could feel her face heating up. Over the past two months, she had been unable to ignore the looks of admiration her boss had bestowed on her. She knew she was attractive. She had learned the power of a soft, innocent stare from those emerald green eyes while still a child, and had perfected the technique on her amused father. She had long ago despaired of avoiding trite masculine compliments about her mouth—kissable and sensual were the usual adjectives. She could pout very enchantingly if necessary.

But she had used no feminine wiles to catch Ron Martens' eye. These were liberated times, and in Nikki's opinion, only opportunists or fools dated their bosses.

She dumped her coat in her office, changed into her shoes, and went down the hall to her boss's office.

"Good morning," he greeted her. He leaned back lazily in his chair and stretched. "I was afraid you'd gotten waylaid in Times Square."

"No, nothing so dramatic," Nikki replied. "I overslept. I'm sorry I'm late."

His expression, which had been warm and smiling, became serious. "Don't look so contrite. The New York *Sun* won't fall apart if Nikki Warren is a little late one day a year. Now, how's your mother doing?"

Nikki responded to the note of concern in his voice by allowing her eyes to warm as she looked at him. He leaned forward across his desk to listen to her, and Nikki instantly regretted even such minor encouragement. She stared over his shoulder, out the window, and replied in an artificially cheerful tone.

"Well, she's getting there, even faster than we'd hoped. She won't make a complete recovery, but she'll be able to do most things. Thank you for asking, Mr. Martens."

"I've told you, make that Ron," he urged intently. "Everybody else does."

Nikki was silent. Ron Martens, who had been working in personnel long enough to recognize frigid withdrawal when he saw it, decided to change the subject to business. "Jack is out sick today, so would you take your lunch early? Say 11:45. I think that's his usual time."

"Will do," Nikki said. "Will that be all?"

"Unfortunately, yes," her boss said with a small smile.

Nikki blushed and hastily walked out. There was steaming coffee and a chocolate doughnut waiting for her on her desk. Her coat hung neatly from the coat-tree. She plopped into her chair and turned to Erika. "Thanks. I'm starved, I didn't have time for breakfast. I guess I'll have to forgive you for that libelous crack about our boss."

"It wasn't libelous, it was true," Erika retorted. "He

looks at you like you're the pot of gold at the end of the rainbow, and you darn well know it."

The two girls got down to work. They began reviewing procedures for hiring telephone solicitors. Erika explained to Nikki that a new Sunday supplement focusing on Connecticut was now being included in papers sold in that state. It was hoped that this would boost circulation in that area, and the current campaign, run out of the New York office, was designed to help recruit additional subscribers.

By 11:45 Nikki was weary of searching for the upbeat, confident personality type who could sell snow in Buffalo—or the *Sun* in Connecticut. She collared her office-mate to tell her that she had to go to lunch early and made her way up to the cafeteria on the tenth floor.

The reasonably priced food at the New York *Sun* cafeteria was one of the best fringe benefits of her job. When Nikki's eyes lit on a delicious looking shrimp salad, she had to admit to herself that it was going to be a nice change from the boring homemade sandwiches she had been bringing in order to save money.

Tray in hand, Nikki looked around the crowded room for someplace to sit; her attention was caught by a familiar laugh that rang out above the noisy buzz of the many simultaneous conversations. She turned to see two friends from the advertising department giggling loudly and waving at her.

"Hi stranger!" This from Beth Browne; her tablemate Sally Corelli called out, "Come sit with us. You'll get some terrific free entertainment."

Nikki sauntered over, pretending to be uninterested. She slid her tray smoothly onto their table and said drily, "Are they running funnier ads now than when I used to work with you? I'll bet they could hear you two all the way to the executive dining room."

Sally put on a haughty face. "If you're going to be that way about it, we won't tell you about the ad that came in this morning, and you'll have to look all through the paper for it yourself."

"Okay." Now Nikki was laughing. "What does it say? Wanted: Actor to play lead role in porno film? Or For Sale: Brooklyn Bridge?"

Beth sniffed, assumed a formal demeanor and began to recite:

"Wanted: Young woman between the ages of 21 and 26 to bear child of unmarried man. Must be attractive, healthy, intelligent. $5000 upon employment, $10,000 upon conception of child, $25,000 upon birth of baby. Send resumé, picture and personal history.

"And then there's a post office box number."

Nikki looked skeptical. "You made that up," she accused.

"Nope," said Sally flatly. "It came on a lawyer's letterhead with a note saying the ad was on behalf of a client. A check was enclosed too. I think it's for real. What I can't understand is—why doesn't this guy just find himself a wife like everyone else does?"

The three of them silently pondered the question for some moments.

"I think he's handicapped, or deformed. Thinks no one would have him," ventured Beth, sympathetic toward the unknown advertiser.

"Or too old and set in his ways to give up his bachelorhood, but he wants an heir to his millions all the same," put in Sally, grinning.

Nikki munched her shrimp salad and said thoughtfully, "Maybe he doesn't like women. Prefers men in his bed, but still wants to be a father. It could all be done in a doctor's office."

Beth burst out laughing. "I should hope so! What woman would want to go to bed with a hunchbacked old man who doesn't like women in the first place?"

"For that amount of money," said Nikki drily, "I'll bet a lot of people would." And silently added, maybe even me.

The conversation turned to shop talk until Nikki, looking at her watch, rose to leave. But as she walked out of the cafeteria, her mind was still on that incredible advertisement. As she rode the crowded elevator back downstairs all she could think was: $40,000! It would pay a lot of medical bills. She should consider the idea rationally, unemotionally. She would never have to meet the man. Her own job would be in no jeopardy; she had known a secretary in the newsroom who had become pregnant during an affair with a married man and elected to have the baby and give it up for adoption. Everybody was totally supportive toward her, and she was back at work two weeks after the baby was born, without having to endure a single sneer or rude joke. After all, this wasn't the Victorian era.

Then Nikki gave herself a mental slap. She must be deranged! Here she was, a little twenty-three-year-old virgin with less experience than most eighteen-year-olds, and she was seriously thinking about having some weirdo's baby? She had to get her mind back to the real world because she had enough problems without adding mental incompetence to the list. With a rueful shake of her head, she entered her office.

Predictably, the rest of the afternoon was hectic. Some of the applicants had to be asked to return the next day. As they gratefully got ready to leave at 4:30, Erika anxiously said to Nikki, "You're looking run-down and we haven't talked for ages. Look, I know your mother wouldn't want you to drive yourself into the ground. You need to have some fun. An old silent comedy is playing up in Yonkers. Ask your mother tonight if she would mind if you come up to my place for dinner and the movie. I'll drive you home afterward."

Although Nikki was grateful for the invitation, she made it a point to spend two or three hours after work with her mother. Nonetheless, she equivocated, 'I'll think about it. I don't like to leave mom alone in the evenings. But thanks."

That evening Nikki arrived at the hospital after grabbing a quick bite to eat at home. It had gotten so that she never even tasted these hasty meals most of the time. She forced herself to wolf down a sandwich or warm up a can of something or other only because she knew she needed nourishment. She could ill afford the ten pounds she had dropped from her already slender frame.

As Nikki walked through the hospital corridors to her mother's room, she reflected on how lucky it had been that if this awful illness had to strike her mother at all, it had happened in the Warrens' store. The store was two blocks from Bronx River Hospital; one of the customers that day was a nurse who had immediately suspected a stroke and had arranged for Mrs. Warren to be taken to the hospital.

Bronx River Hospital was affiliated with an excellent medical school and was considered one of the finest hospitals in New York City. The doctors on the staff were first-rate, and even if Nikki sometimes needed a translation of their medical lingo she had found them to be honest and sympathetic. Her mother's room, in the new Hannah Cragun Pavilion, was cheerful and clean.

Her mother's face, though drooping slightly on the right side, registered concern along with her usual smile of welcome as Nikki entered her room. Mrs. Warren's speech had not been affected too severely by the stroke, and had improved rapidly. Now she spoke quite clearly, but slowly, as if the words needed to be carefully thought out in advance.

She took her daughter's hand, squeezed it, and told her, "Nicola—please—eat—rest. You look—awful."

"Oh, Mom!" Nikki objected gently, "I know I look tired, I had a busy day today. But surely I don't look *awful!*"

"Yes." The tone was insistent.

Nikki tried to wriggle out of the accusation. "Yes, I look tired, or yes, I don't?" she laughed.

"You look—*awful.*" Mrs. Warren was utterly implacable.

Nikki gave up and sighed. "You're the second person who's scolded me today. First Erika, now you. I think I'm going to have to listen."

"Good," Mrs. Warren said with satisfaction.

"Then you wouldn't mind if I had dinner with Erika and went to a movie with her tomorrow? She lives up in Yonkers. She takes the train back and forth, but she has a car and she said she'd drive me home."

Nikki realized she saw relief in her mother's eyes, and Pam Warren said sadly, "I worry. If you—were married—"

"Pamela Beale Warren, are you trying to blackmail me?" Nikki asked accusingly. "Because if you are, sick or not sick, it won't work."

But Nikki knew that to her mother she was still in many ways the pony-tailed little girl whose tears she had wiped away and whose cuts she had bandaged. Mrs. Warren could not accept that Nikki was all grown up now and well able to take care of herself, even if at the moment her haggard appearance indicated that she wasn't doing a very good job of it. Still, she didn't need some overbearing male to shepherd her around.

This was not the time to engage in feminist polemics, however, and Nikki sought to reassure her mother. "When the right guy comes along, I'll want to get married. But there's no one right now. Don't worry. I promise that you'll have a grandchild one day."

Then, seeing that her mother was tired, she kissed her a loving good night and added, "I'll go home to get some rest. That should make you happy. See you Wednesday."

Nikki's own remark—that her mother would have a grandchild one day—came back to mind as she half-ran, half-walked the few blocks through the chill winter air to her home. A sudden impulse made her stop at the corner drugstore and purchase a copy of the next morning's *Sun*. She had recollected her lunch-hour conversation and wanted to find out if that absurd ad was really in the paper.

Upstairs, she went into her apartment to search eagerly through the bulky paper. She soon located the ad which Sally, Beth and she had laughed over—it was printed in unusually large type and stood out boldly. Telling herself that it would be amusing to show it to her mother on Wednesday, she clipped it out and put it into her wallet.

The next night, as Nikki sat convulsed with laughter in the movie theater watching the "Little Tramp," she silently blessed Erika for suggesting the outing.

They had decided to go see the film first and eat afterward. As they left the movie, Nikki thanked her friend for her concern.

"I guess I needed a good laugh. And if I'm as thin as you say, all the popcorn I ate should be good for at least five pounds. You know, Erika, I've felt the world closing in on me, and I've forgotten that life goes on and we all muddle through somehow. Hell, if the tramp can make it, then so can I!" she laughed.

"So philosophical!" said Erika sardonically. "I only hope it lasts. By tomorrow you'll probably be agonizing over your financial problems again. You need someone to share the burden."

Nikki answered warmly, "You are, Erika. It means a lot to me that I can talk to you."

Erika regarded her a little shyly. "I know you're an only child, but don't you have any relatives you can turn to? Or family friends?"

Nikki shook her head and explained, "Mom has a few close friends who come to see her during the day. They look in at the store for her from time to time also. But they don't have any extra money lying around. My mother's sister and brother-in-law are retired down in Florida with a son in medical school in California. That's expensive. And my father's half-brother lives in Washington state. I've met him exactly once, years ago. I don't think he and dad got along, even as kids. So that leaves me."

Erika decided it was time to lighten the conversation. "You should tell your troubles to Ron Martens," she joked. "He'd be thrilled to listen."

"Oh no, Erika, not that again!" moaned Nikki, who realized she was being teased and was quite willing to go along.

"Well, he obviously likes you. If you gave him any encouragement—" Erika persisted impishly.

"Encouragement!" yelped Nikki. "This morning I gave him a friendly look and he practically scaled the desk between us. If I batted my eyelashes at him—"

"He'd propose," Erika finished. "He's in the market

for the second Mrs. Martens. If he looked at me the way he looks at you, I'd be ecstatic."

"Well I'm not!" came the definite reply. "I like him well enough, but not as a boyfriend. It's a bad idea to mix business with pleasure."

Erika shook her head and gave a mock grimace. "Okay, I get the message. You don't want to discuss Ron Martens. Come on, I'll fill you full of chicken soup. That cures everything."

"Very funny," said Nikki. But she laughed.

The two girls had their chicken soup. And steak, half a bottle of wine and apple pie with ice cream. Soon, mellow and relaxed, Nikki talked emotionally about her parents, how close they had been, the shock of her father's death from a heart attack after his automobile had been wrecked by an uninsured drunk driver, her empathy with her mother. Erika in turn entertained Nikki with tales of her own childhood and adolescence.

Nikki lay in bed later that night and thought about her evening. She realized that her sides still ached with laughter. Through all the talking and giggling and even a few misty moments, she and Erika had become much closer. Instinctively she knew that Erika Berger was someone she could talk to, could trust. Nikki had told her only a fraction of what she felt, but even that was enough to lighten her mood. For the first time in weeks she fell asleep almost immediately, smiling to herself.

During the next few days the office was like the Bellevue Emergency Room on New Year's Eve. The phones rang constantly; soon the personnel department stopped setting up any more interviews. Some of the would-be solicitors were ultimately placed elsewhere. In spite of how busy she was, Nikki was happy. Her appetite had improved. Mrs. Warren was doing much better; she was obviously relieved to see her daughter in such good spirits. Even the prospect of a weekend at the store—going over inventory, deciding what to order, checking the books, paying the bills—did not dent Nikki's good mood.

At 4:45 a.m. on Friday, the phone in the Warrens' apartment rang. Nikki grabbed it, instantly awake and filled with panic. At that hour of the morning she knew it could only be bad news. It was. Mrs. Warren had suffered a second stroke, a deep male voice told her. It was not life-threatening, but perhaps Nikki would care to come to the hospital as soon as possible?

The resident on duty told her that he felt this "episode," as he called it, was less dangerous than the first, but would surely retard her mother's progress considerably. Tests were being run now.

Nikki prowled the halls impatiently, and finally went down to the coffee shop for something to eat. Only the thought of her mother worrying about her persuaded her to force down the toast and tea she had ordered. After a few hours, since there was nothing she could do and no information to be gleaned, Nikki left and went to work. The hospital knew where to reach her if necessary; the doctor assured her that they would know more by that evening.

In the middle of the morning a call came in from a woman who introduced herself as Miss Sands from the hospital business office. They were terribly sorry to bother Nikki. This must be a difficult time for her. The nurses were so upset about her mother's latest setback. But she owed $2200 and now it looked like the bill would go much higher. How was she going to pay?

Nikki, striving to sound calm and confident, assured Miss Sands that she would take care of the matter. But she thought to herself, how indeed?

Somehow she managed to get through the rest of the day. Her boss had wanted to send her home, but she refused. She knew it was better for her to keep busy.

Every Friday Nikki deposited her paycheck in a bank near the office, keeping only as much cash as she needed for the week. As she stuffed two twenty dollar bills into her wallet, she picked up a piece of paper that had fallen out. It was the ad she had clipped four days before. She had completely forgotten about it.

The news at the hospital was not encouraging. Although there was no major new damage, Mrs. Warren would have to begin all over again.

Nikki's mother was alert and very determined to recover. Nikki was amazed anew at her courage. When the subject of money arose, Nikki convincingly lied that there was no problem, since the savings were still available. Mrs. Warren had no idea of what her treatment was costing. Hospital expenses had risen sharply in recent years and Pam Warren still thought in terms of her husband's hospitalization ten years before, following his first minor heart attack.

That night, depressed and drained after a long day at work and an evening spent sitting at her mother's bedside, Nikki withdrew the ad from her wallet and reread it. She wrote a brief personal history and found a recent photo of herself in her mother's picture album. Her emotions were so battered that she carried out these tasks stiffly, as if she were in a trance.

She mechanically attached a copy of her up-dated resumé, stuffed everything into a manila envelope, addressed and stamped it, and dropped it into a mail chute outside her apartment door. Only after the envelope was actually out of her hands did the potential consequences of what she had done strike her.

For a moment she was horrified that she had answered the ad at all. How could she have imagined that she would go through with such a thing? Then she relaxed. She had hardly agreed to it. She would probably never hear from the man or his attorney. Suddenly she giggled, then laughed hysterically. The tensions of the last twenty hours seemed to dissipate.

Chapter 2

During the next week, Nikki sublet her apartment. Two rental payments were no longer possible, even though there was some income from her mother's store to help. The money saved would pay for a few days' hospital care. She sold some jewelry which had belonged to her father's mother, although she hated to do so; she also cashed in three bonds she owned. She knew her mother would disapprove, but what choice was there? Erika insisted on loaning her some money; altogether she raised about half of what the hospital demanded. It seemed to satisfy them for the time being.

She came to appreciate what a nice group of people she worked with. Everyone was so supportive. Her boss made sure that she was assigned routine, easy tasks, or interviewing, which she continued to love.

Mrs. Warren again astounded her doctors by her unyielding persistence in getting well; she was being taken out of bed already and asking them when she would be able to try walking alone. She refused to be depressed by her latest setback. In light of her mother's cheerfulness and faith, Nikki found herself becoming more optimistic.

Her failure to attend to the store the previous weekend nagged at her conscience. At the time it had seemed more important to be with her mother, but she promised herself that she would catch up on the ordering during the weekend to come. Otherwise they would start running out of merchandise in a few weeks, and that was surely no way to show a profit.

It was a cloudy Thursday afternoon. The intercom in Nikki's office buzzed. "Call for you, Nikki. Line three."

Nikki clicked down the button and said, "Nicola Warren here. May I help you?" She heard an efficient-sounding female voice intone, "Please hold the line for Mr. Morris, Miss Warren."

Mr. Morris, thought Nikki. Who's he?

A kindly masculine voice came on the phone. "Miss Warren, my name is Charles Morris. I'm an attorney. I represent the client whose ad you answered. I'd like to talk to you in person, if I might. Perhaps you would be good enough to stop by my office this afternoon after work?"

Nikki felt mildly nauseated. She could not believe this was happening just when she had begun to think that her life was manageable again. Of course she was going to tell him that she had changed her mind about the whole thing.

What came out was an unsure stutter. "Uh, yes. I mean, no. That is, I *did* answer the ad, but I'm not—I'm no longer interested."

"Why don't you drop over and explain it to me in person, Miss Warren? Nobody's going to strong-arm you," Charles Morris answered smoothly.

"There's nothing to explain. It was an impulse—it's a long story—" Nikki fumbled for the proper words.

"Perhaps you would prefer it if I came to see you?" the lawyer asked sympathetically.

"No!" Nikki gasped, horrified. Why wouldn't the man understand that she was sorry she had ever answered the blasted advertisement? If she had not feared that this Charles Morris would materialize on her doorstep, she would have hung up on him.

"Your office or your apartment?" he persisted calmly.

Nikki felt that she had no choice but to see him, to make him understand that her involvement in this was a much regretted mistake. Resigned, she heard herself say, "I'll come to your office. Where and what time?"

It had just begun to snow as Nikki left the *Sun* building. She had planned to walk to the attorney's office—it would not be much slower than the crowded uptown bus. But the heavy, wet flakes were turning to water as they landed, and she had no desire to arrive at this unpleasant interview disheveled and soaked. Feeling guilty because of the expense, she hailed a taxicab and then tried to relax as the driver threaded his way across town and north, rapidly changing lanes and running red lights as if maneuvering in New York City traffic were some sort of giant sporting event.

The law offices of Morris, Clayborne and Chase were in an elegant old building on Fifth Avenue. How convenient for his clients, thought Nikki drolly. If they have time to kill before their appointments, they can pop into Tiffany's and pick up a few extra diamonds.

She walked down the softly carpeted hallway toward a set of double doors, trying to make up her mind whether or not to go in. Surely if she simply failed to show up, the lawyer would not pursue the matter. But as she hesitated in front of the beautifully carved dark wood, a gray-haired man opened one of the doors and smiled at her.

"Miss Warren?"

Nikki muttered a faintly strangled "Yes." What on earth was she doing here?

"I'm Charles Morris," he told her. "We have a security system in the building—a television camera. I recognized you from the photograph you sent. If I may say so, it didn't do you justice."

He continued suavely, "Come in, come in. Everyone else has left for the day. I thought it would be more comfortable for you to talk to me without a lot of other people around."

He ushered her into a plush office furnished with antiques and motioned vaguely toward a blue velvet couch.

"How about a glass of sherry or wine?" He walked over to a cabinet which housed a small refrigerator, took out a bottle of white wine, and held it up for her to see. White-faced, she nodded. She needed something to get her through this interview, because Charles Morris struck her as the type of man who bulldozed people until he got what he wanted.

Nikki sipped the wine and thought, Nicola Warren, don't have too much of this stuff or you're liable to do something even more stupid than what you've already done. She glanced nervously at the man next to her on the couch, wondering how to begin.

"A bit more relaxed now?" the attorney asked her.

"Not really," she replied truthfully.

"I think I'll go ahead anyway. I'm a bit pressed for time.

"Your application was impressive, Miss Warren. Your grades in college show you've got brains and I can see you've got beauty. I'll take your word on the health part of it. You've got a good job so you're not starving. I can only assume that for some reason, you need the money very badly."

"Very perceptive." Nikki attempted a smile, failed miserably, and drank some more wine.

"Lawyers are supposed to be able to figure things out. Do you want to tell me about it?" he asked in a kindly voice.

"No, not really," Nikki repeated. Her glass was empty. She held it out, asking, "May I have some more wine, please?"

"Sure. Help yourself."

Nikki poured it with a badly trembling hand, and almost knocked over her glass when the lawyer abruptly asked, "When can you meet my client?"

"Meet him?" Nikki yelped. "Even if I agreed—and I haven't—I'd never—I mean, I didn't think he'd actually want to meet the woman—" Oh lord, she thought, this gets worse and worse.

Charles Morris ignored her distraught expression. His voice became soothing, coaxing. "Miss Warren. I've already told my client about you, and he's very anxious to meet you. Of course he wants to make the final decision himself. I just screened out the kooks. But you'd be surprised how many serious applications there were."

Nikki stared out the window. It was nearly dark. Suddenly her curiosity overcame her common sense.

"Mr. Morris, can I ask you a question?"

"Sure. Go ahead."

"I think I'm entitled to know something about your client," she plunged in. "That is, assuming I was interested in his proposition. I—I don't understand why he has to hire a female body—to have his child. Why doesn't he just get married and do things the normal way?"

"I'm afraid I'm not at liberty to answer personal questions about him. I'm sure you can understand that. But you can ask him anything you like when you see him. I'm sure he'll be more than happy to discuss the matter with you. If nothing else, you'll enjoy meeting him. Go talk to him. You can always say no."

Well, he has a point there, thought Nikki. She was beginning to feel a little lightheaded from the wine. Why not go? It would be good for a laugh.

Aloud she said, "Maybe I will. Do I have your word I won't be kidnapped?"

The attorney laughed. "Kidnapped may be the right word for it. You're sure you're really twenty-three? Because you don't look it."

His teasing encouraged Nikki to relax. Her eyes sparkled as she replied drily, "Yes, I'm twenty-three. Or so my mother tells me."

"Good." There was no question that Charles Morris was well satisfied with his half-hour's work. "My client's building is on Sixth Avenue. It's not that far from the *Sun* building." The attorney handed her a piece of his stationery with an address written on it.

"I know he's out of town until Monday, Miss Warren. So how about if I set it up for Tuesday at five?"

"You understand that there's no way I'm actually going to *do* this, Mr. Morris," Nikki said in what she hoped was a firm voice.

Charles Morris grinned at her. "I'd say that's my client's problem. My job was only to persuade you to go to his office," he admitted disarmingly. "Although I have to tell you in all honesty that he's made up his mind to see you. If you had refused to go to him . . ." An expressive shrug filled in the rest.

"In that case," Nikki drawled, "your client has

something in common with you! Who should I ask for?"

"Just go to Suite 3001," Charles Morris directed. "I *do* have your word that you'll be there?"

"Why not? I think I'm curious to meet this mysterious client of yours. I've led too sheltered a life."

As she spoke the words, Nikki realized that it was the truth. Normally she was quiet, steady and logical. Today she felt a bit reckless, and it wasn't really the wine. Her mother's illness had brought home to her the uncertainties of life. Why not do something wild and unpredictable and bizarre for once in her twenty-three years? After all, she told herself, this lawyer knows I'm not about to have a baby. I'll just meet this client of his and find out what it's all about, then report back to Beth and Sally. We'll be laughing about it for months.

She shook hands with Charles Morris and smiled good-bye. He's really very nice, she mused. He wouldn't be representing somebody disreputable; his client must be a sincere man with perfectly sensible reasons for placing the ad.

The five days until Tuesday dragged by with frustrating slowness. Things had quieted down at the office so Nikki did not have the constant pressure of work to occupy her thoughts. She spent altogether too much time wondering about the identity of "the mystery man." She told no one, not even Erika, what she had done. Her acquaintances would have found it strikingly out of character for rational, reliable Nikki, and she had no wish to explain her motivations. She came to look on the whole matter as a whimsical adventure, a lark. At least for five whole days her mind had been off her financial problems, and when she pictured her coming encounter with Charles Morris's client she invariably experienced a sense of anticipation, amusement.

Ron Martens continued to ask about her mother's condition. He was really quite sweet, Nikki said to herself, and with his sandy hair and gray eyes, rather good-looking. All she would have to do is smile at him, touch his sleeve, and he would be dragging her off to have dinner with him. But she refrained from smiling too brilliantly, and from touching, and as a result was dragged nowhere.

Though her mother's second stroke had not proved as damaging as the first one, Nikki's supposition that it would still add many weeks of treatment and thousands of dollars to the hospital bill was correct. Nikki had discussed the matter thoroughly with her mother's doctors, and they told her to count on at least six to eight weeks more of hospital care, even if her mother continued to improve at the same rapid rate.

Nikki had promised herself that she would tackle the store that weekend, so she spent Saturday and Sunday checking out inventory, filling out order forms, and paying bills. She was relieved when she had these chores out of the way.

On Tuesday afternoon Nikki called the hospital's business office to explain that she made a good salary and to promise that all bills would be settled eventually. She offered to produce personal and credit references if they wished. She was determined to come to some agreement with the hospital because she did not want to be even remotely tempted to accept "mystery man's" ridiculous proposition. She recognized that he might turn out to be as persuasive as his attorney, God forbid! Nikki would simply go and meet him, explain she had changed her mind, and file this away as a story to tell her children one day.

But the voice on the other end of the line interrupted

her explanations with "Oh yes, Miss Warren. There's no problem with your account. The matter's all settled."

Puzzled, Nikki thanked her and hung up. She supposed that the hospital had already checked her credit rating, found it to be excellent, and decided to wait for payment.

Tuesday was crisp and clear. Nikki enjoyed the brisk, fifteen block walk uptown. The address on Sixth Avenue turned out to be a brand new building. She screwed up her expressive face at the unoriginality of the design.

But she had to admit that the inside was lovely. The two-story lobby contained a huge abstract stainless steel sculpture, a sparkling fountain with intricate streams of water playing over and among its metal garden, and dozens of trees and plants in large redwood containers.

The listing of offices next to the elevators announced: "Central-Atlantic Industries Building." Suite 3001 was "Executive Offices, CAI, Inc."

So Sally was right, Nikki decided. Of course she had heard of CAI—it was one of those modern octopus corporations with tentacles in every pond. Her "mystery man" must be a retired official or board member, never married, who now wanted an heir. She felt almost guilty about going upstairs under false pretenses. She hated to waste the time of some old man. But she had promised. . . .

She stepped out of the elevator into a large reception area—all glass tables, modern sectional furniture, plants. Several hallways led away from this central area. At a sleek, bare desk looking more decorative than functional sat a staggeringly beautiful redhead.

"My name is Nicola Warren. I have an appointment for 5:00." And if you ask with whom, Nikki thought to

herself, I'm going to tell you that your guess is as good
as mine.

"Oh yes," the redhead simpered. "I was asked to
wait for you, Miss Warren. Go right down the corridor
behind me. The office is at the end. It's open; you can
go right in. He'll be back in a moment." Nikki thanked
the woman and walked down the long hallway.

For some reason, though Nikki had felt mildly
amused until then, she began to tremble as she walked
into that office. She had never been in a room like this:
oriental carpets, what she guessed were Chinese and
Indian antiques, pre-Columbian art, expensively upholstered furniture.

God, he must be loaded, she thought.

She walked by a starkly modern steel and mahogany
desk which was devoid of any papers. Either the man
was a workaholic who compulsively cleared away every
scrap of paperwork, or he was a figurehead with little to
do. She guessed the latter—perhaps the unofficially
retired Chairman of the Board.

She went over to a window overlooking Sixth
Avenue to watch the traffic below. She wasn't aware
that anyone had entered the room until a deep,
mocking voice came from behind her.

"Admiring the view? Or were you thinking of
jumping?"

She jerked around, then simply gaped. He was in his
early thirties, she judged, tall (at least 6' 3"), dark hair
which needed to be cut, blue eyes. And handsome—the
best looking man she had ever seen, except possibly on
a movie screen. His features were classic, except for a
slightly crooked nose which looked like it had been
broken. His thin mouth wore a decidedly cynical smile
at the moment. Nikki knew that he looked somehow
familiar, but she could not place him. He couldn't
possibly be her "mystery man."

His next words corrected that mistaken notion.

"Finished vetting me? But then I guess you're entitled," he drawled. "It's not every day you meet the future father of your child."

Nikki tried to hide her disconcertion but missed by a mile. Her voice quavered as she asked softly, "Who are you? Your lawyer—Mr. Morris—he wouldn't tell me your name."

He ignored her question, but walked over to her with lithe, easy grace and said, "You'll let me help you off with your coat, Nikki."

She didn't care for his informality and liked his ordering her about even less, but nonetheless politely thanked him for his help. He seemed too formidable to tangle with just yet.

"Not at all," he replied. "I merely wanted a better look at what's underneath." He proceeded to examine her as if she were an amoeba under a microscope. She had worn a new dress—a cream-colored wool sheath—with gold jewelry and brown boots. At home the outfit had made her feel older, more sophisticated. But the manner in which he was scrutinizing her made her think that she might as well have walked in naked. She was sure the man was visualizing her that way. His gaze had begun at her booted feet, traveled up to her midsection, and lingered on her mouth.

What little composure she had left deserted her utterly when he said lazily, "Not bad. A bit scrawny—for having children, that is." He paused, then added, "Oh yes. You wanted to know my name. It's Michael Cragun. I'm disappointed that you don't recognize me." But he hardly sounded it.

Nikki *did* recognize him—now. She had seen the picture of CAI President Michael Cragun in both the business and social pages of the paper. One of New York's most eligible bachelors. But she was hardly about to give him the satisfaction of telling him so. Instead she remarked as coldly as she knew how, "Are you always so arrogant? Or is this a special performance for my benefit?"

He smiled—a slow, devastating smile that stopped short of his stony blue eyes. "Don't flatter yourself, Nikki. Sit over there." He pointed to a comfortable-looking couch.

Nikki resented this second order as much as she had resented the first one, but she obeyed because her legs felt like they were composed of petroleum jelly. Somehow she had completely lost the initiative here, if indeed she had ever truly had it in the first place.

"Mr. Morris must have explained that I—" she began breathlessly, only to be interrupted by Michael Cragun's curt, "Leave Charlie out of this. His part in it is finished."

Okay, Nikki agreed silently, I'll spell it out for you. Aloud, she announced rebelliously, "I won't do it, you know."

"We'll see," he said with lazy confidence, and turned his back to her.

Goaded by his rudeness, Nikki decided to be equally rude. "Why do you need some stranger to have your kids?" she challenged. "I know you're rich and famous and all that—there must be women standing in line for the rare privilege of meeting you. Why don't you do us all a favor and get married like normal people do?"

He ignored her outburst. With a cold look in his eyes he walked toward her carrying a tray holding a bottle of wine and two half-full glasses. She took the glass he held out and sipped, avoiding his gaze.

"Maybe that will calm you down," he mocked. "Now. Let's get the relationship straight. *I'm* hiring *you. I* ask the questions."

Oh no you don't, she wanted to say, because I'm getting out of here.

But somehow absolutely nothing came out. And then he remarked conversationally, "I suppose you've had your share of lovers, otherwise you wouldn't be here. Have you ever been pregnant?"

His matter-of-fact tone infuriated her. "That's none of your damn business," Nikki said stiffly, even though her sense of humor compelled her to recognize that, under the circumstances, it most certainly was his business.

She was relieved when he sat down, even though he unfortunately chose to join her on the couch. He was so tall, it put her on the defensive to have him towering over her, looking down at her as if she were a recalcitrant child. He reached out his hand and with taunting deliberateness put it under her chin to turn her face toward his. She flinched as he touched her, pushed his hand away, and glared at him.

He said dangerously, "You'll answer the question, Nikki."

"Okay," she spat out. "No. I have no track record as a brood mare."

He threw his head back and laughed. "One point for you, Nikki! But don't count on winning the game. You're playing against an undefeated champion."

My God! she thought, his arrogance was incredible! She said, "I don't count on winning, Mr. Cragun, because I'm not even playing."

Something about his expression, so cool and bored, as if she weren't worth listening to, provoked her. "I answered your stupid advertisement as a joke," she fumed. "My friends and I wanted to see who would put in something so completely outlandish. Beth said you were handicapped or deformed, and you're not, except possibly for a warped mind. Sally thought you'd be an old man who wanted an heir. And I—" She hesitated. Telling him her opinion was obviously going too far.

"Yes? Do go on. I find this fascinating."

His mocking tone pushed Nikki past caution. "In spite of outward appearances," she said, her voice dripping with sarcasm, "I conclude I was right. You won't have a child in the normal way because women aren't in your game plan, to use your trite metaphor. You only play with men."

Feeling quite triumphant, Nikki got up to leave. Michael Cragun grabbed her arm and pulled her slowly back down. His mouth closed over hers, intentionally rough. He kissed her in a passionless, insulting manner, and when she tried to keep her lips clenched tightly shut, he forced her mouth open and relentlessly explored it with his tongue. She attempted to push him away, but he was far too strong. His hands bit into her shoulders and upper arms, tightening when she struggled. Finally the hateful kiss was over, but he continued to hold her loosely by one arm.

Nikki's whole body felt violated, her lips bruised and swollen, her mouth aching. She was furious, trembling, her nostrils flaring with anger. "Let go of my arm. I'm getting out of here."

He lay back, still holding onto her arm, and put his feet up on the cocktail table in front of them. He seemed completely relaxed. The kiss had certainly not aroused him; it seemed to have had no effect whatsoever. He inquired lazily, "What about your mother?"

His question shocked Nikki. She had told no one her real reason for answering the advertisement. She was trying to come up with some coherent response when he spoke again.

"Who's going to pay her medical bills?"

Nikki took a deep breath. At least she had an answer

to that! "I don't know how you found out about that,
but it's all settled. The hospital will wait."

"You're right," he said conversationally. "It *is*
settled, because I paid what you owe and told them to
bill me in the future. By the way, I've had your mother
moved to a private room. And I arranged to call in my
own specialist, although I'm told her doctors are very
good. When the time comes, your mother will be
admitted to the best rehab center in New York. I assure
you it isn't that easy to get a patient into the March
Institute."

Nikki stared at him, totally stunned. She truly had no
idea of what to do next. The idea that someone could
casually take over her responsibilities without her
permission or knowledge was surrealistic. She was
suddenly frightened of Michael Cragun.

He released her arm, got up, walked over to his desk.
The heavy manila folder he removed from the top
drawer was dumped on the table in front of her.

"Go on," he ordered curtly. "Open it up."

She did so. She was desperately trying to keep her
body from shaking. When she skimmed the contents, it
was all she could do not to throw up. Inside were copies
of school records, interviews with former teachers,
classmates and colleagues, a complete financial report,
details about her mother's illness, and much more.

Michael Cragun had sprawled into a chair and was
watching her indifferently. "You didn't really suppose,"
he said, "that I would hire you without knowing all
about you."

Nikki was her mother's daughter. She might feel sick
with apprehension, but she also knew something about
fighting back. Her face was burning, but she succeeded
in keeping her voice icy cold.

"I don't know how you got all this information. Some
of it is confidential, and I'm appalled that people I

know were so willing to talk about me. But I suppose someone in your position can do things mere mortals like me can't. In any event, I don't care. My mother can do without your fancy doctors, and I'd walk the streets before I'd take one dime of your money."

This little speech had no apparent effect on him. Nikki told herself that she had only imagined a fleeting glint of admiration in his eyes.

"I told you not to count on winning the game, Nikki." His smile reflected his amusement at her attempt to strike back at him. There was a pause, and then he said in a nonchalant voice, "Your mother's in a new section of the hospital, Nikki. You remember what it's called?"

The change of subject rattled her. She snapped out, "What the hell difference does it make—" then abruptly clamped her mouth shut. She recalled the name: the Hannah Cragun Pavilion.

He went on in a detached tone of voice, "My father made a fortune in plastics. Central-Atlantic Industries was built around his original business, but now we own everything from a movie studio to hotels to an auto parts manufacturing firm, and much more. Of course we're a public corporation now, but the family still retains a controlling interest. Hannah Cragun Pavilion is named for my grandmother. My father's way of thanking the hospital for some very excellent care during heart surgery a few years ago. He's on their board of directors, by the way."

He sipped some wine. "If you refuse my assistance, you're going to owe a great deal of money. And my family has a certain amount of influence at that hospital."

Nikki studied the floor, mesmerized by his last words. This couldn't really be happening to her. When would she wake up? She asked softly, "Mr. Cragun, are you threatening to have my mother thrown out of Bronx River Hospital?"

"I don't threaten," he stated bluntly. Nikki wondered what else one would call it—blackmail perhaps?

"I simply want to make you aware of—the score of the game, shall we say? The lease on your mother's store runs out in two months. If someone else were to buy that building, the rent could be raised or the lease not renewed. As for your job—I have a few friends at the *Sun.*"

"How impressive." Nikki seized on the one threat she could answer. "I belong to the union. I can't be fired just because you—"

"What makes you think I don't have friends in the union also?" he interrupted.

Nikki tried again. "You seem to forget that I *do* work for a newspaper. If the reporters I know ever got hold of this story—"

"It would never be published," he finished for her. "I told you, I have friends at the *Sun.*"

Now Nikki was silent. Michael Cragun had wealth and power, but surely no one could really do all the things he had threatened. And if he could, why go to all the trouble? There were other women—extremely eager ones, if the gossip columns could be believed.

Her thoughts were interrupted when he got up from his chair and sat next to her on the couch. "Nikki."

She stiffened at his nearness and refused to look at him.

"Nicola."

At the use of her real name, her eyes filled with tears. Only her mother ever called her that.

Michael sensed that she was weakening, and he pressed home his advantage by saying gently, "Be a good sport. Concede the game and let me take care of things for you. Admit that it would be nice to stop worrying about money, to let someone else deal with your mother's business."

He took out his wallet. "Here. A check for $5000. Above the medical expenses, of course; I'll pay for those. Take it."

Nikki's eyes rested on the check, made out to her, but she silently shook her head and forced herself to control the panic in her mind and shuddering of her body. She tried the one ploy that had never failed her where men were concerned. Winsome, vulnerable green eyes were turned up full blast and aimed at Michael Cragun. Her full lower lip was pushed out just a fraction to tremble with hurt and indignation. She was about to launch into a heartrending plea for her freedom when he broke into the first uncynical grin she had seen on his face.

"Look at me like that, Miss Warren, and you'll find yourself pregnant within the next fifteen minutes," he laughed.

It was not the reaction she had counted on. She was sure her cheeks were stained pink as she mumbled, "No thanks. Not in fifteen minutes and not ever."

His expression became hard. Nikki had the feeling that every motion Michael Cragun made, every word he spoke, was planned. He was no doubt a consummate actor, but she had no stomach for applauding the performance.

He said in an uncompromising voice, "I've gone to a lot of trouble to find out all about you. I've decided that you're going to have my child. I'm not going to sit here and waste my valuable time arguing with you about it. If you think your life's been difficult so far, I promise it's going to be a complete misery unless you agree."

"Why? Why me?" she burst out. "What difference—"

"I told you," he cut in, "I ask the questions and you answer them. We're going to the hospital to visit your mother, and then I'm taking you out to dinner. And I swear to God I'll *carry* you out if you try to refuse!"

Nikki had no intention of refusing. Anything was better than sitting in this intimidating office along with the overpowering Michael Cragun. She snatched up her coat to forestall any attempt he might make to help her on with it, and scurried out ahead of him.

As they left the now-deserted building, a smiling middle-aged man emerged from a shiny black Cadillac limousine and held the rear door open for them. His steely hair was cropped close to his head, making him look like a Marine; he wore a white shirt and dark pants in lieu of a uniform.

Michael Cragun grinned and shook his head. "Sorry we took so long, Henry. Say hello to Nikki Warren. Nikki, Henry Merola, my chauffeur, right-hand man, and husband of the best cook in Manhattan."

She nodded politely in response to the chauffeur's cheerful "Hi there." The gray-haired man couldn't help it if he worked for a total lunatic.

Not a word was exchanged between Nikki and Michael Cragun during the trip to the hospital. Nikki was convinced that he was crazy. Maybe the pressure of business and his social schedule had combined to unhinge his mind. She thought it prudent to go along with him for tonight, and then to refuse to see him again.

About halfway to the Bronx, Henry Merola and his boss started to discuss proposed CAI stock acquisitions. Apparently Henry was an avid and unusually successful speculator. To Nikki's way of thinking, this kind of talk was properly conducted in the confidential atmosphere of executive board rooms, not in front of a hostile third party. Although she would never dream of selling the information she was overhearing, it made her uncomfortable to have to listen. She was relieved when the Cadillac pulled up in front of the hospital.

Several people smiled hello to the tall man at Nikki's side as they walked through the busy corridors. As they left the elevator, Nikki automatically turned right. When Michael put his hand under her arm to stop her, she was so startled that she jabbed him viciously in the stomach with her elbow. Then she backed away and looked nervously up at him, half-expecting him to hit her back.

Instead he pointed toward the left. "I told you your mother was in a private room now. It's down this way. And for God's sake, relax. I can hardly ravish you in the middle of Bronx River Hospital."

She followed him down the hall and tried to figure out how she was going to explain him to her mother. As just a friend, she supposed. His first words to Pamela Warren caught her off guard.

"Mrs. Warren." He spoke slowly, gently, as Nikki kissed her mother hello. It was not a tone of voice he had favored Nikki with. "You're looking better than you did this afternoon. Can I be so conceited as to think you had the nurse do your hair for my benefit?"

Pam Warren answered, not without difficulty, "Absolutely."

I don't believe this, Nikki told herself. They're like old friends.

Pam Warren looked at her daughter. "Why have—you—been hiding him?"

"I haven't, mom." She tried to sound calm, even though her emotions were oscillating wildly between incredulity and anger. "We've only known each other—a short time."

Nikki stiffened as Michael's arm was draped over her shoulder. Really! The man deserved an Academy Award! He smiled at her mother, then with his back toward the bed looked down at Nikki with mocking eyes.

"Darling, your mother knows that I wouldn't be paying the bills unless I were more than a casual friend."

Oh yes, said Nikki silently, you're more than *that*. But our relationship isn't what mom thinks it is.

Michael Cragun turned back to Mrs. Warren. "I spoke to the man in my office whom I mentioned to you. He said he'd enjoy running your store for a while. And it would save Nikki from having to do all the inventory checks and ordering and bookkeeping." He smiled charmingly at the admiring woman in the bed. "She's gotten herself very run-down in the past few weeks. I intend to take care of that. I want her healthy."

I bet you do, Nikki thought. Only the best for Michael Junior.

She could tell that Michael was being deliberately provocative—phrasing things to have double meanings—and swore that she wouldn't let him rile her. Still, she was increasingly furious that she hadn't known about his visit this afternoon.

How had he gotten in? Mrs. Warren's visitors were restricted, and his name wasn't on the list. But then, he seemed to know half the staff.

Nikki's musings were interrupted by Michael questioning Mrs. Warren, "Tell me, has your daughter always been so stubbornly independent?"

Pam Warren smiled back, "Always." She laughed when Michael commented, "I'm going to have to take her in hand, I can see that." As if to underscore his intention, he pulled her closer, his arm around her shoulder tightening possessively.

He must have sensed by the way Nikki tensed with rage that he had pushed her to the limit. He loosened his hold on Nikki to tell Mrs. Warren about the specialist he had engaged; the rest of the visit passed uneventfully. Nikki was only too content to let him take over the conversation; she doubted her ability to discuss anything rationally by now. After half an hour, Michael announced that he was taking Nikki out to dinner "to fatten her up." She glared at him.

Michael directed Henry to a seafood restaurant on New York's City Island. The fishing village atmosphere of the island made it seem more like a New England seaside town than part of aggressively urban New York. The restaurant was justly famous for its excellent fish, and crowded even on a Tuesday night as a result. Nikki, anxious to escape, was about to inform Michael that she was too hungry to wait when the headwaiter appeared, greeted Michael by name, and led them to a quiet table for two in a secluded corner.

Nikki refused to look at her menu, coldly ignoring her escort's polite inquiry as to what she wished to eat. He had dragged her here after mercilessly taunting her at the hospital; she had no intention of being an amusing dinner companion for such an insufferable man.

Michael merely shrugged and ordered for her. She told herself that she would not give him the satisfaction of eating one single bite of the food he had paid for, but the lobster looked so succulent that she was unable to resist.

Nikki continued to sit stonily, determined to be as unpleasant as possible without provoking him to the point of retaliation. But she had not reckoned with Michael's even stronger determination to charm her out of her black mood.

His initial questions about her job were answered with brusque monosyllables. But then Michael launched into a monologue about his own experiences in hiring people, amusingly classifying job applicants according to their quirks and personalities. Nikki had to smile; she recognized every type he described. After that there was no returning to her state of unresponsive aloofness. Nikki discovered that Michael Cragun was a very good listener; he seemed so genuinely interested in everything she said that at first she suspected blatant manipulation of her emotions. And then she simply didn't care. His coaxing charm was so persuasive that she loosened up in spite of her earlier vow not to do so.

They exchanged humorous childhood stories and discovered a mutual love for the theater. Nikki found that Michael had also majored in economics in college, and did not bother to bristle at his teasing, "Of course *I* already knew that we shared the same major." Like all proper sons of corporate magnates, however, he had subsequently gone on to Harvard Business School.

As Michael walked her to her door Nikki was floating, and not from the mere half glass of wine she had been careful to limit herself to. She had never spent such an enjoyable evening with a man. Her brain conveniently repressed that awful interview in his office and the deliberate provocation at the hospital; all she knew was that she was bowled over by his charm and intelligence. She realized with a jolt: I want him to kiss me good night.

She looked up at him, trying to sound cool even though her heart was pounding and her body felt hot all over. "Thank you for dinner. I enjoyed it, Michael."

"So did I. But why thank me? Self-interest, remember?" he asked. "I want you to be well-fed and content, just like all my other livestock."

But his tone was teasing and he was smiling. Nikki blushed, finding it impossible to take offense at his words, and fished around in her purse for her keys. He took them from her, opened the double-locked door, and held her hand to press the key case back into her palm. Nikki was aware that he kept her hand in his for longer than was strictly necessary. When he lifted her chin and looked down into her eyes, it was all she could do to keep from reaching up and stroking his face. "I'll be in touch," he said softly. "Henry will be driving you back and forth to work for the next few days. Good night, Nikki." He trailed his mouth lightly down her neck, winked at her, and left.

Sleep did not come easily that night. Nikki had always tried to be honest with herself. She knew that she had been so aroused, so pliant, that if he had chosen to follow her into her open apartment, she would have made no demur. And if he had led her into one of the bedrooms. . . . She refused to let her thoughts go further. The fact was, he hadn't.

He was obviously very experienced when it came to charming the opposite sex. His name had been linked with many different women: beautiful, wealthy, famous women. There was that governor's estranged wife, and the actress who was last season's television sensation; the daughter of an automotive tycoon and a European princess.

God, she thought, I must have seemed dull by comparison. Childish, ignorant, and boring.

A man like Michael Cragun wasn't remotely within her frame of experience. She hadn't traveled all over the world, or hobnobbed with scores of famous people, or indulged in numerous lighthearted affairs.

And, she told herself, that's why I fit his requirements. He isn't interested in making love to me, because if he were, he would have tried something tonight. I gave him a clear enough invitation. I must have struck him as a besotted schoolgirl; he only wants me as a factory for growing a child. The women he's linked with would find the idea of wasting nine months to have a baby ridiculous. She resolved to guard both her emotions and her actions.

She could not begin to fathom what his next move might be. The uncertainty was both frightening and exciting. Then she checked such treacherous thoughts. She might be inexperienced, but she was no fool. She would never be stupid enough to permit herself to fall in love with Michael Cragun.

Eventually, she almost convinced herself of this and fell asleep.

Chapter 3

Nikki heard nothing from Michael Cragun during the next few days. Disappointment mingled with relief. As per Michael's instructions, Henry Merola took her to work each morning and drove her home each evening. Nikki contemplated refusing to ride in the chauffeured

Cadillac, but the thought of Michael Cragun appearing like some avenging devil to bodily drag her into the car quickly made her abandon the idea of resistance.

The first morning, she asked Henry to leave her two blocks from the *Sun* building and to meet her at the same spot at 4:40. She had no wish to answer leering questions about why she was riding in a limousine these days. If this arrangement struck Henry as unusual, he was mute on the subject.

After the topic of the weather had been exhausted, Nikki had asked Henry indifferently, "Have you worked for Mr. Cragun for a long time?"

He replied in a friendly but guarded tone, "Five years, Miss Warren."

Well, deliberated Nikki, what have I got to lose? She was intensely curious about the handsome, wealthy Mr. Cragun. So she ventured, "Has he ever been engaged? Or serious about anybody?"

Now the chauffeur's tone became the slightest bit cool. "Well now, Miss Warren, I don't discuss the boss. I'm happy to talk about the weather some more, or how the Mets will do this summer. But don't ask me about the boss."

Nikki had been politely put in her place, and she knew it. Thereafter she avoided the subject of Michael Cragun, and she and Henry held casual, non-controversial conversations during her rides downtown and back. One day she casually remarked, "Your driving is sure an improvement over the subway. I think I could easily become very spoiled by all of this."

"I guess that's the idea," he answered cryptically.

All day Wednesday, Nikki had successfully resisted the impulse to go down to the newspaper morgue, the room where old clippings were filed, for more information on Michael Cragun. After all, what was he to her but a manipulative male who for some bizarre reason of his own had chosen her for a target? Not for anything would she admit to unbridled curiosity.

By Thursday, however, she had managed to convince herself that it was only sensible strategy to learn everything she could about her enemy.

She pored over the clippings that night. The *Sun* did not indulge in printing gossip, but she learned that the Cragun family was active in a broad range of charities. Michael had taken over the presidency of CAI, Inc. four years before, at the age of twenty-eight, after his father had undergone open-heart surgery. The business news recorded the names of companies CAI had acquired or sold; the list was lengthy. There was nothing personal about the man, but the amount of wealth he controlled and his contacts in the social and business communities were staggering—and intimidating.

It was impossible for Nikki to reconcile the arrogant, implacable businessman who had kissed her with such brutal thoroughness in his office with the charming, teasing man-about-town who had trailed his mouth down her neck in the hallway of her apartment house. The first man frightened her badly. It was important to Nikki to shape her own future and handle her responsibilities effectively. He had entered her life like a tractor, intent on mowing down her autonomy and self-respect.

The second man was something else again. Nikki admitted that she had a crush on him. She suspected that his attentiveness had been a subtle method of insuring her compliance with his plans. Yet the way she felt when he looked at her or touched her made her *want* to comply. And Nikki came to realize that fighting the second Michael Cragun would be much more difficult than fighting the first.

Friday afternoon Ron Martens poked his head into Nikki's office. She felt someone's eyes on her, looked up, and smiled. "Did you want to see me?"

"Yes," he told her. "Come down to my office."

A minute later she was seated across from him as he explained, "I just got a call from the publisher's secretary. The honor of your presence is requested on the top floor." He shot her a questioning look. "Do you have any idea what it's all about?"

"No." Nikki was apprehensive. "No, honestly, I don't."

"Neither do I. So go ahead up. And take your coat with you. You won't be coming back."

Nikki paled. "What do you mean? That I'm going to be fired?"

He rushed to reassure her. "No, no. Only that P.D.'s secretary asked if I could do without you for the rest of the day. I could hardly say no. So have a nice weekend, Nikki. I'll see you Monday."

"Thanks. You too." Nikki's voice was distant. Back in her office she left a note for Erika and threw her purse and coat over her arm. Making her way through a plushly carpeted hallway, she came to an elegant set of doors marked, "Peter Welsh Delavan, Publisher." The receptionist directed her to the left. She followed the small hallway and timidly knocked at the office door.

A male voice called out, "Come on in, Nikki." She slowly opened the door to see Peter Delavan sprawling back in an expensive leather chair with his feet resting up on his desk. Michael Cragun was leaning negligently against the side of the desk. The two of them were laughing over something.

So, Nikki thought, this was Michael's "friend" at the *Sun*. You couldn't go any higher. But then, she should have guessed immediately. The man obviously knew half of New York, and the other half was probably beneath his exalted notice.

The two men rose as she came in. Michael walked over to her, kissed her lingeringly on the mouth, and said, "Nikki, meet your boss, Peter Welsh Delavan. Pete, a lady whose talents are wasted on your rag of a paper, Nicola Warren."

The publisher laughed. "Michael, my friend, you only say that because you have other plans for her. How do you do, Nikki?"

She moved away to shake his outstretched hand; to her consternation Michael placed an arm firmly around her waist and walked forward along with her. What on earth did he mean, "other plans"? Michael wouldn't have said anything. Or would he? Peter Delavan was studying her so intently. . . .

Michael released his hold on her to take her coat from her arm and help her on with it. She had no choice but to let him do so, even though her body was reacting in the most annoying fashion to his hands, which came to rest on her shoulders, pulling her back against him. Then he turned his attention toward Peter Delavan. "Thanks for letting the help out early, Pete."

"Sure. Any time. And let me compliment you on your selection. I know she's got brains because she works for the *Sun. And* she's a stunner, Mike. You know, it's a great story. When you're ready to make an announcement, let me know."

"Will do." Good-byes were said, but Nikki scarcely knew what amenities she was mouthing. She was in shock. Michael must know she would never agree to the whole ridiculous scheme. How could he casually tell his friend about his plans for fatherhood? Nikki seldom became angry—at least she hadn't until she met the infuriating Michael Cragun—but now her temper began to rise.

Later she hardly remembered being led out a back entrance into a private hallway toward the publisher's personal elevator. In the hall, she turned on her companion, in a fine rage.

"You beast! How could you tell him about—" she hissed.

Michael interrupted, "I didn't tell him anything."

This bored, slightly annoyed interjection drove Nikki beyond endurance. She threw down her purse, cut him off with "You're a liar!" and slapped him as hard as she could across his left cheek. Her satisfaction at the red marks her fingers left on his face was short-lived. Michael immediately grabbed her wrist to pull her arm down; Nikki kicked him viciously in the shin to try to free herself. The next moment she found herself pinned between the wall and his hard body, her two hands forced high above her head by his right hand. She struggled, twisting, wriggling, kicking at him again. She was too out of breath to speak. Her attempts to escape proved futile, he was too strong, his body was pressed solidly against hers, holding her without effort.

His voice was tight with anger as he said, "Okay, that's it. The first time, in the hospital, I let you get away with it. But no more. If you were a child I'd turn you over my knee and spank you. But since you're not . . ."

The memory of that insolent, bruising kiss in his office flashed into Nikki's mind. She prepared herself for a repeat performance, determined not to give him the satisfaction of resistance which both of them knew would be a waste of her energy.

His method of punishment was far more subtle. He backed away to slowly unbutton her coat and pull down the zipper of her sweater. Then he took her face gently in his hand and began to kiss her neck. She held her body rigid, hating him. His mouth traveled to her ear, nibbled it, then trailed to her mouth. She kept it closed, her teeth grinding together. If that was his game, she'd be damned if she were going to respond.

He continued to kiss her, sweetly, all around her mouth, then chewed softly on her lower lip, tracing the outline of her mouth with his tongue. His thumb had dropped to her hip to begin a drugging massage.

Nikki's resistance melted. All she knew was that she was as aroused now as she had been angry before. He was brushing his mouth lightly over hers when she realized that her whole body was shaking. The temperature in the hallway seemed to have risen about ten degrees. She could not endure any more of this tender assault. The only thing that mattered was that he should kiss her properly. Not in anger, but in passion.

She closed her eyes, relaxed her mouth, and felt him draw slightly away.

"Oh, no. Not yet," he said softly. "First you apologize for slapping and kicking."

Nikki opened her green eyes wide, looked into his glowing blue ones, and whispered, "I'm sorry, Michael." She was in no fit state to argue.

"And now you ask me to forgive you," he ordered in a perfectly controlled voice.

She was silent. But then his mouth began caressing her neck again, his body moving intimately against hers. She tried to turn her mouth up to his, but he refused to let her. Finally, her temples throbbing and her breathing ragged, she forgot whatever pride she had and choked out, "Yes. Please." When he made no move toward her, she asked achingly, "Why are you teasing me like this, Michael?"

He stared at her liquid eyes and parted lips for a moment, then muttered, "No more." There were no

tender preliminary explorations. He took her mouth
with deep, hungry kisses. He had released her arms;
she stood on tiptoe, arms now around his waist,
pressing her body close to his. Never in her life had she
responded so totally. Michael slipped his hand under
her sweater to deftly unhook her bra; he moved his
fingers lightly over one breast. Although for a moment
Nikki was stunned by the expert passion of his lovemaking, she soon began to kiss him back, at first
hesitantly, then with increasing confidence and passion
as she realized the effect she was having on him.

She was bereft when Michael slowly pulled away and
took his mouth from hers. She swayed, watched
dazedly as a broad smile chased across his face, only to
be replaced by a look of utter arrogance. He ordered,
"Let's go!"

Nikki, still trembling from his lovemaking, could not
believe he would treat her in such a manner. She
whispered, "Wh—what?"

"You heard me! Move!" The tone was harsh.

Nikki felt small, humiliated, and terribly vulnerable.
He had treated her with such exquisite gentleness and
she thought that he had soon become as aroused as she
was, that he liked her at least a little bit. Was the whole
thing some sort of game to him? Did it amuse him to
treat her as if he'd bought her for an hour? Nikki
looked down to hide the tears that filled her eyes, and
shook her head.

"I said we're going." His tone was distant, bored.

Nikki was unable to answer or look up. She lacked
the sophistication to laugh about this interlude or to
disguise the expression of bewilderment and pain on her
face. Michael muttered an impatient "Oh, for God's
sake!" then picked up her purse, and all but dragged
her down the hall and into the elevator.

The black Cadillac was waiting in front of the building. Nikki was too nervous and off balance to think of escape. She did not return the affable Henry's "Hi, Nikki" for fear that traitorous tears would spill down her cheeks. She would not make even more of a fool of herself by actually crying.

Michael directed his chauffeur in a laconic voice, "Take us up to Nikki's apartment, Henry." Then he made himself comfortable on the seat, pulling some papers out of a leather briefcase as if absolutely nothing out of the ordinary had just happened.

Nikki sat huddled next to the window behind Henry, as far from Michael as she could possibly get, and thought dejectedly, so he's taking me home. For some reason she felt like a naughty child who was being sent to her room. She wished fervently that she were alone, free to cry her heart out. When she felt Michael's hand against her hair Nikki jerked around, hurt and mistrust in her dilated green eyes. His casual assumption that he could touch her whenever he pleased was awful. "Why don't you just leave me alone?" she said bitterly. "Get your kicks somewhere else!"

His answer astonished her. He announced in a voice that brooked no rebellion: "I got you out of work early because I'm taking you to Florida for the weekend. My parents are down there for the winter; I want to see them and discuss some business with them. The weather's been beautiful—it will do you good to get some sunshine."

Nikki doubted that he was serious. He made love to her one minute, snarled at her the next, and now he wanted to take her to Florida to meet his parents? What kind of Byzantine plot was this?

She attacked, "You're crazy. I'm not going anywhere with you."

There was a disgusted sigh, followed by a smile and a lazily teasing answer: "Doesn't the prospect appeal to you? Sitting on a quiet beach, soaking up the sun?"

His seductive smile punctured her defenses.

Michael continued, "I talked to your mother this morning. She thought it was a good idea. You know how she worries about you. You should call her before we leave to say good-bye to her, Nikki."

Nikki tried to sound as cool as *he* seemed. "Really?" she asked. "And did you tell her what our relationship is? Seeing that you plan to announce it to the world in the New York *Sun*?"

He began to laugh. "If you'd given me half a chance to explain before you hauled off and belted me in Pete's hallway, I'd have told you that the announcement he was referring to was of our engagement. I told him a romantic story about how we met in the hospital coffee shop. He thinks you're going to be my fiancée."

Nikki was incredulous. His fiancée? Why would he say that? Granted he wanted to get her out of work early, but he didn't have to go that far!

She told herself that she deserved some explanations, and was damn well going to demand them. She had a right to know Michael Cragun's plans for her. But her tone when she spoke was more defensive than forceful.

"I don't understand you. You say you want me to have your child, but I can't believe you really mean it. You—you seem to think I'm some kind of wind-up doll. Does it—does it give you a sense of power to make me do what you want? To—to turn me on when I'm angry?" She glanced out the window at the traffic, then went on painfully.

"I suppose you know everything there is to know about making love. You're—you're very good-looking, and smooth, and you may think of me as a toy, but I

happen to be human. I admit that I respond to you, but
that still doesn't mean I like you. And—and to tell your
friend we're going to be engaged—I suppose that's your
twisted sense of humor, but I wouldn't marry you,
Mister Cragun, even if you went down on one knee and
begged!" She lifted her chin and stared at him challeng-
ingly, proud that she had finished with a flourish.

"A very impassioned speech, *Miss* Warren," he
mocked coldly. "You'll no doubt be relieved to know
that a proposal of marriage is hardly imminent." He
paused and suddenly grinned at her. "Now. You are
coming to Florida with me. As to my evil intentions
toward your underweight body, I'll decide in a few
weeks, when you look less like a scarecrow. And,
honey," he finished huskily, "if you don't want me to
touch you, don't provoke me. *I* happen to be human
also."

Michael looked out over Nikki's shoulder at the
highway. The traffic was literally bumper to bumper
and moving at a crawl.

"If we ever get off the damn Cross Bronx Express-
way, we are going to your apartment where you will
pack a suitcase. Is that clear?"

Nikki felt completely drained. She had no strength
left with which to fight him. Her emotions had been
stirred up to begin with, and his baffling combination of
highhanded arrogance and careless charm had left her
totally confused. Anyway, his parents would be there
as chaperones and a weekend in the sun was scarcely
ten years at hard labor.

Nevertheless, she felt compelled to show him that
some small streak of independence remained. She
turned to face him, raised her hand to give a mocking
salute, and in a perfect imitation of his lazy drawl said,
"Yes sir, General Cragun. Sir."

His satisfied reply: "You're finally learning, Private Warren."

Michael announced that he would come upstairs while Nikki packed, and once there, stretched out on her bed, made himself completely at home. The intimacy of the situation disturbed her, but then Michael Cragun had no doubt been in so many women's bedrooms that they apparently weren't very arousing to him anymore. He watched with a detached expression as she filled her suitcase. Nikki made no protest. She was afraid that if she did, he would respond in a manner that, she ruefully admitted, she lacked the experience to handle.

When she pulled out an old black one-piece swimsuit, Michael was quick to object. His voice registered distaste as he asked, "Is that thing all you own?"

Nikki nodded, keeping her back toward him, and tossed the swimsuit into the suitcase.

She heard Michael begin to get off the bed. "I think I'll have a look, if you don't mind," he said nonchalantly.

Nikki hastily produced a pink bikini, turned to hold it up for his inspection, and said sourly, "Will this do, Your Majesty?"

"First a general, then a king. I'm moving up in the world." He studied the bathing suit, as if picturing it on her body. "Hmm. Yes, it'll do very well, even in your present emaciated state."

It was foolish of her to let these teasing remarks about her thinness disturb her, but they stung nonetheless. She noisily threw a few more garments into the suitcase and dramatically banged it shut. "Will you excuse me while I change into something cooler?" she asked with frigid politeness.

Much to her surprise, Michael made no sarcastic

rejoinder, but simply slid gracefully off the bed and walked out, flicking the door closed as he left. Nikki changed into a navy and white pantsuit and entered the living room to find him engrossed in a serious phone conversation with her mother. He had the sheer gall to stand there and solemnly assure Pam Warren that he would take good care of Nikki. He held out the phone to her, his eyes registering delight at her outraged countenance.

She in turn endured her mother's gushing praise of Michael, and gritted her teeth when her mother insisted that she promise to "do as Michael says—he's so busy with work, but he wants to look after you." It was a relief to end the conversation.

As she said good-bye to her mother, Michael fetched her suitcase and took her keys from her. He locked up the apartment and they went down to the car. Michael flipped the keys to the chauffeur, telling Nikki that if Mrs. Warren needed anything from her apartment while they were in Florida, Henry would be happy to bring it to her. It was all arranged.

The dinner flight from Kennedy Airport to Ft. Lauderdale Airport in Florida was about to taxi over to the line of planes waiting to take off when Michael and Nikki approached the gate. The blond airline attendant on duty smiled at Michael, "Another thirty seconds and you wouldn't have made it, Mr. Cragun."

As they took their seats, Nikki needled nastily, "Of course, they would have held the plane if they'd known *you* were coming."

"Probably," he answered tonelessly.

"But then," she went on, courting danger but unable to help herself, "I'm surprised you fly on a commercial flight, even if you do go first-class. Doesn't a big, important tycoon like you have his very own jet for these little jaunts?"

He turned to her, and for once his voice wasn't bored, or taunting, or drawling, or seductive, or curt.

"The company has a plane, yes. But this isn't a business trip, Nikki. I'd like to have a relaxing weekend with my parents. I can't do that if you keep baiting me. I know you're angry about the situation you've gotten yourself into, or that you think I've forced you into. But for forty-eight hours, can you please forget it? Let's call a truce and enjoy the sunshine, hmm?"

Nikki did not trust this new sincerity of his for a moment. She crushed the impulse to respond to it. "What happens after forty-eight hours?" she snapped.

Michael stretched his hands up above his head, yawned, and then rubbed his eyes. "We'll talk Sunday night. Frankly, I've had a hell of a week, Nikki. If you don't mind, I'm going to try to sleep for a while." Without waiting for her reply, he clicked out the overhead light, put his seat back, and closed his eyes.

Within two minutes his breathing was deep and regular. Asleep, he looked boyish, even innocent. It was hard to believe that he was an aggressive business-man who, it would seem, single-mindedly pursued what he wanted. In this case—me—Nikki thought.

Michael slept through dinner, but Nikki surprised herself with her appetite for the fresh fruit, filet mignon, and most of all, the champagne that the flight attendant kept pouring.

As her nerves unwound, Nikki tried to recall every-thing that Michael had ever said to her. She wondered what he was really like—how he behaved when he wasn't playing games or acting a role. Apparently he was close to his family, and the idea of watching him with his parents, of getting to know him, was appealing.

She would have to do her best to pretend that he was simply a friend of hers, a man who had invited her for the weekend to meet his parents. Then it struck her that Michael's parents would classify her as another one of their son's bedmates, and she felt cheapened and ashamed.

Michael woke five minutes before they landed, and was half-asleep and still yawning when they left the plane. He seemed oblivious to the fact that he had placed an arm possessively around Nikki's shoulders.

Nikki's initial impression of Mr. and Mrs. Cragun was that they could indeed have been the parents of any of her real friends. She had expected to be met at the airport by another chauffeured limousine. Instead, just as her suitcase was coming through into the baggage claim area, Ann and Jake Cragun pulled up in a two-year-old white Chrysler. Jake was tall with gray streaks in his dark hair—dignified, handsome, smiling. Ann was a willowy blonde who looked elegant even in the casual tee shirt and slacks she wore. Nikki saw immediately where Michael had gotten his good looks—his parents made a strikingly attractive couple.

Their son waved as they entered the terminal and met them halfway, enfolding both of them in a warm hug. All three walked back arm-in-arm to where Nikki was standing, holding her suitcase. He took it from her, telling his parents, "Mom, dad, I want you to meet a friend of mine, Nikki Warren. Nikki, my parents Ann and Jake Cragun." Ann immediately told Nikki to call them by their first names—"All of Michael's friends do," she smiled.

The conversation on the way to their home was about the bitter weather in the Northeast, how the Yankees were doing in spring training, and about Michael's sister Melanie, whose eight-year-old twins had come

down with chicken pox. Nikki was content to sit in the
back seat next to Mrs. Cragun, relieved that no
questions were probingly directed toward her.

The Craguns' home was a four bedroom ranch house
backing on a private beach; it was situated on the
narrow island of Key Biscayne. The living room was
furnished in a comfortable, unostentatious manner in
shades of lime green, yellow, off-white, and beige.
Although Nikki was certain that everything was expensive, the house had a warm, welcoming feel. Her
attention was held by the baby grand Steinway that
stood in one corner of the living room. She was about
to utter the usual compliment about the beauty of her
hostess's home when Ann Cragun spoke to her son.

"Michael, why don't you take Nikki's suitcase down
to your room?" She smiled at Nikki. "Nikki, you can
unpack while I fix some coffee and cake."

Nikki, rooted to the elegantly carpeted floor with
embarrassment, looked uncertainly over at Michael,
who was making no attempt to check his laughter.

"Mom," Michael said, a big smile still on his face, "I
can see that I've gotten you well trained. But Nikki and
I won't be sharing a room. Why don't I put her in
Lauren's room? She can watch the sun rise over the
ocean tomorrow morning."

Jake Cragun stared at Nikki with new interest, and
muttered, "Well, this is a switch." His wife shot him a
quelling look and turned to apologize to Nikki.

"I *am* sorry, Nikki. And I think I won't say another
word, or I'm going to get into worse trouble than I'm
already in!"

Nikki's laugh, though forced, dispelled Mrs.
Cragun's chagrin. Michael picked up the suitcase and
led Nikki away. When they were alone he told her,
"Mom stopped arguing with me about sleeping arrangements when I was a kid of twenty-three." He

smiled at her, that devastatingly attractive smile of his. "Just your age, in fact. But I'm sure she's gratified to learn that there's still virtue in this jaded world."

On the plane Nikki had vowed not to bait him, but his statement was unfair provocation. "What makes you so sure?" she taunted. "You once told me I'd had my share of lovers. How come you've changed your mind?"

"You forget, I know all about you. I know you were so busy with school and helping your mother that you had very little time for a social life. I know you've never had a serious love affair."

She reddened as he went on softly, "And I've made love to you—or started to. If I wasn't sure before, I am now. By the way," he added enigmatically, "if you weren't such a babe-in-the-woods you wouldn't be here now."

Nikki unpacked slowly, trying to decipher his last cryptic comment. Did he mean that only an untarnished virgin was suitable as the mother of his proposed child? The idea was so insolent that it was comical. More likely, her naive resistance appealed to his jaded sense of humor.

With a sigh, she went to join the Craguns in the brightly colored breakfast room adjoining the kitchen. They were in the middle of a confidential business conversation. She hesitated before sitting down, saying lightly, "This sounds like a top-secret meeting. I think I'd better go read a book."

Michael motioned her into a chair. "Come on, sit down. You didn't object when Henry and I were discussing stock transactions. I know you'd never make it as an industrial spy, you're too honest. I'm not worried." Nikki didn't know whether to be insulted or complimented.

It was past 1:30 a.m. when Mrs. Cragun announced

that if her husband and son were planning to play golf early the next morning, they had all better get some rest.

Nikki sleepily undressed, feeling as high as a kite. She had enjoyed the stimulating conversation and the evening's informal atmosphere. She had never pictured millionaires lounging in the kitchen talking business, only in formal living rooms exchanging inane gossip. She was flattered by the friendliness of Michael's parents and pleased at the way she had been included, trusted. The discussion had touched on many extremely confidential areas; Michael must have a high opinion of her honesty and integrity, even if he considered her hopelessly inexperienced and childish in many ways. She was so wound up that she was sure that she would never fall asleep. But the hypnotically soothing sound of the surf crashing against the shore soon had its way.

Chapter 4

Nikki woke to find that everyone else had gone out. Michael had referred to this as Lauren's room; Nikki assumed that he meant the niece whom his mother had spoken of as Laurie. A quick inspection of the closet revealed clothes that were almost large enough to fit Nikki but much too youthfully trendy for her taste. The daughter of Michael's older sister must be a teenager.

Nikki drew aside the drapes to look out the picture window at the sea, which glistened brightly through the trees in the morning sunshine. She wandered into the kitchen, where she found a note from Mrs. Cragun stuck onto the refrigerator door with masking tape.

Dear Nikki,

You were sound asleep; we didn't want to wake you. Jake and Michael are off playing golf; I have to go to a meeting but I'll be back for lunch. Make yourself some breakfast—fresh-squeezed orange juice in the fridge, compliments of my son. There's suntan oil in your medicine chest and towels under the sink, so take a swim. Feel at home!

Ann

P. S. Don't swim in the ocean. The Portuguese men-of-war are biting!

The kidney-shaped swimming pool was located beyond a screened-in flagstone patio. The thought of lying in the sun proved irresistible to Nikki, and soon she was lounging in a deck chair in her bikini, a glass of iced orange juice nearby.

Later on, shaking off her sybaritic sloth, she took a walk on the beach. A palm tree lined path through the vegetation led to the ocean.

Ann Cragun returned at 12:30, accompanied by a tall Cuban woman whom she introduced as Mrs. Muñoz. The Craguns did not have a full-time cook; Mrs. Muñoz came in to prepare the food for their dinner parties.

Michael's mother had found Nikki stretched out in a deck chair, half asleep. She had given a gentle tug on the girl's hair to get her attention, then smiled. "I can see you've had a relaxing morning. I'm glad, because Michael is worried about your health. But I'm starved, and Maria hates to be disturbed when she works in the kitchen. Get dressed and we'll go out to eat something fattening for lunch."

"Are you having people over tonight, then?" Nikki was surprised by the appearance of the formidable looking Mrs. Muñoz. Michael had mentioned nothing about a dinner party; she felt that she was intruding on a family occasion.

Ann Cragun asked in a strange voice, "Michael didn't tell you anything about it?" When Nikki shook her head, Ann explained, "You see, we'd already invited several of our friends when Michael phoned Thursday to say he was coming. But he told me to go ahead. He's polite enough to pretend he likes socializing with us old folks when he's down here, which isn't often enough. Did you bring a long dress with you?"

"No, I'm afraid I didn't," Nikki told her hostess. "I packed rather quickly, since Michael gave me about ten minutes notice. But I'd feel out of place at a family party."

"Don't be silly, dear. Michael wants to show you off—of course you'll come," Ann insisted. "I'd lend you one of my dresses or one of Melanie's, but they would be too large. So while we're in Bal Harbor, we'll go shopping. There's a Saks Fifth Avenue there, and you can use your charge card."

Bal Harbor, like Key Biscayne, was a wealthy area near Miami, in South Florida. Nikki was not in the habit of cruising exclusive department stores unless it were to window-shop. It was a bit discomfiting to have to inform her hostess, "I'm sorry, Mrs. Cragun, but I don't have a charge account there."

"Please don't worry about it, Nikki," Anne insisted. "I'll put it on my account and we'll settle up later."

The two of them chatted companionably over a mouth-watering salad of fresh fruits and vegetables and a dessert of chocolate torte with whipped cream.

Over lunch, Nikki had airily decided that Michael had gotten her into this absurd situation and he could darn well pay the bills for it as well, and the two women were soon strolling into Saks.

It was a new and welcome experience for Nikki to be ushered into a luxurious private dressing room while clothes were brought to her. Ann, as Nikki was now calling her, was obviously a well-known and valued customer. Nikki had heard that the saleswomen in such places were sometimes dreadfully snobbish, but they practically kowtowed to Ann Cragun. Dress after dress was produced for their inspection.

Ann persuaded Nikki to take a backless, halter-topped silk dress, all sunshine colors of yellow and orange. She announced that Nikki *must* have a silk flower to wear in her hair (". . . and do leave it down, dear") and some high-heeled sandals (". . . so we won't have to shorten the dress"). Their shopping finally accomplished, they drove home.

The two women arrived to find father and son in swimsuits, drinking beer and lazing around the pool. Nikki could not prevent herself from thinking that if Michael were handsome in clothing, he was even more so without it. His body was that of an athlete in training: lean, muscled, hard. He had powerful thighs that looked like those of a runner; a broad chest covered with a smattering of dark hair; a flat stomach. It was all Nikki could do to tear her eyes away from him. Their eyes met; she knew he had noticed her admiring reaction, and she flushed. Nikki decided that she had to get away from Michael's disturbing influence. She asked her hostess if she could make a phone call to her aunt and uncle in Ft. Lauderdale.

"Of course," replied Ann. "But you'll be seeing them in a few hours." She took in Nikki's bewildered expression. "I didn't want to mention it before, because I thought it must be a surprise. Michael asked us to invite them when he called Thursday night. I'm sorry if I spoiled anything, Michael."

"No," he said with a shrug. "It wasn't a surprise. I just didn't think of mentioning it."

Without stopping to consider, Nikki began to ask Michael how he had known her aunt and uncle's name and telephone number. When she paused in mid-sentence, a wicked grin lit up his face.

"I know all about you, remember, honey?" He stared at her, enjoying her discomfort.

In addition to Nikki's aunt and uncle, the Craguns had invited four couples to dinner. Nikki emerged from her room aware that she looked sensual and lovely; her attractiveness gave her confidence. Her hair, as Ann had suggested, flowed silkily over her shoulders and back, which were covered only with a light sunburn. An orange silk flower was pinned to one side of her center part, giving her a vaguely Polynesian look.

Michael and his parents were standing in the living room when she walked in; he merely commented "very nice" in an off hand manner and coolly offered her a Margarita.

Nikki's aunt and uncle, Sarah and Jack Newhouse, arrived first, to be quickly followed by the other guests. Two were neighbors; there were friends Mrs. Cragun had met through her charity work and the president of a citrus bottling company and his wife. The company was owned by CAI; Nikki wondered if there were any area where CAI didn't have an interest.

The Newhouses immediately drew Nikki aside and began to question her. As they sipped their Margaritas, Sarah Newhouse said with concern, "Tell us how your mother's been doing since you wrote."

Nikki explained in meticulous detail about the second stroke; she told them exactly how her mother had been affected physically; what therapeutic steps would be taken and the prognosis after rehabilitative treatment.

Her Uncle Jack said guiltily, "I wish we could help with the money, Nikki. But with Terry in medical school—"

Money was the last thing Nikki wanted to discuss. She hastened to assure her uncle and aunt that everything was working out, and prayed that they would believe her.

Her Aunt Sarah looked dubious. "If I said I'd been surprised by Mrs. Cragun's call, it would be a total understatement. Since when do you socialize with people like their son? I thought actresses and heiresses were more in his line."

Her voice took on a mother hen tone. "Now, Nikki, I feel responsible for you with your mother so ill. I think you should tell me if this man is paying the bills. Have you agreed to become his mistress in return for—favors granted?"

From Nikki's point of view, Michael Cragun could not have picked a better moment to appear and announce smoothly, "I'm going to have to steal your niece away. Everyone wants to meet the beautiful brunette I've brought to Florida with me."

Instead he led her out through the sliding glass doors to the patio, and told her drily, "You looked like you needed rescuing; that scene in the corner reminded me of the Spanish Inquisition, with Nikki Warren as victim." He stroked her hair. "You know, playing the white knight on a charger isn't usually my thing, but since the damsel in distress is looking so beautiful tonight—"

Nikki could have told him that *he* had created the situation from which she needed rescuing, but on this warm, romantic evening, she had no wish to begin that argument again. She dropped a small curtsy and said huskily, "Thank you, kind sir. The damsel in distress appreciates both your well-timed rescue and your gallant compliment." The dress and the Margaritas made her flirtatious; she flashed him her most brilliant smile.

He gazed at her mouth for a moment, then glanced over her head into the living room, where ten pairs of eyes were watching the two of them with undisguised interest. "I think we should go inside," he told her as he firmly took her arm. "It's getting cold out here."

"Don't be silly, Michael. It's still in the 80's. Don't you like being out here with me?" Nikki asked in her most provocative voice.

He turned her around to face the windows between the patio and the living room. Hands on her shoulders, he drawled, "I have to admit that even I have never made love to a woman in front of an audience. But if you'd care to be the first?"

Feeling completely safe, Nikki took advantage of this perfect opportunity to get back at him for his earlier teasing. She asked mischievously, "Would you really?" Her fingers began to play with his tie, then traveled up to brush his lips.

Michael captured her hand and pressed a kiss into her palm. He warned softly, "Behave yourself, little witch. I made myself a promise to be a gentleman with you this weekend, but in a moment I'm going to break it. I've told you—don't provoke me."

"Yes, Michael," Nikki dimpled triumphantly, and sashayed inside ahead of him.

Ann Cragun soon summoned her guests to enjoy the delicious Mexican dinner produced by Maria Muñoz. Nikki, seated between the Craguns' neighbor Alex Bryte and her Uncle Jack, engaged in bantering small talk. From time to time she glanced at Michael, but he seemed to have no interest in her.

After dinner the well-fed assemblage retired to the living room to talk and sip brandy or amaretto. Nikki followed Michael to the couch, where she sat down close beside him, her thigh touching his. Her intention was to punish him for his indifference during dinner. Unfortunately, though his closeness had a dizzying effect on her senses, the reverse did not appear to be the case.

After fifteen minutes, however, he got up to click off his father's sound system. "With all due respect to your latest toy, dad, I'd prefer some live music. Nikki, play something for us. We wouldn't," he added with a grin, "want all those years of music lessons to go to waste."

Nikki's temper flared. Not only did she resent being ordered around—the man did not seem to know the word please—it was more than a little annoying to have someone know everything about you when you knew so little about *him*. Her eyes sought Michael's and she could tell that her peeved reaction was not only obvious to him but that he found it entertaining. One more point for him.

Yet it was a beautiful piano. She was anxious to try it out. "Are show tunes all right?" she asked no one in particular. "I'm not in a classical mood." Her suggestion was met with enthusiasm; she started to play the romantic theme from a current Broadway hit. Then she silently groaned and for the rest of the evening she tried to avoid tender love songs.

The Craguns' guests left late, well past midnight. Ann and Jake walked arm-in-arm toward their room; Nikki suspected their hasty withdrawal was a tactful ploy to leave their son alone with her.

It was left to Michael to turn off the lights in the kitchen and living room; Nikki stood by the piano and watched. When he passed her to go to his room, she whispered, "Don't you want to kiss me good night?"

He stopped, turned. "No." Then informed her coolly, "Don't start something you won't want to finish. Go to sleep." And taking his own advice, he stalked down the hall to his room, slamming the door behind him.

The next morning over breakfast, Jake Cragun asked Nikki if she had ever visited the Florida Everglades. When she told him no, that this was her first trip to Florida, he promptly suggested to his son that she would enjoy going.

Nikki had been unable to repress the memory of her actions on the patio and in the living room the previous evening, and by breakfast time she was appalled by her own behavior. She had all but propositioned Michael, and she was certain he was fed up with her on-again, off-again manner. His good morning to her had certainly been grumpy enough; thereafter she avoided his eyes. Anyway, trekking through a wildlife preserve seemed completely out of character for such a city man.

She answered Jake hurriedly, "Oh no, perhaps some other time. Michael came down here to see you and to relax."

"Can Michael decide what Michael wants to do?" he said lazily. "I wouldn't mind going. The last time I was down there was during a drought, and I saw almost nothing. How fast can you get ready?"

"Are you sure?"

"Nikki, I said I was. Now get moving!"

She ceased arguing. "Is fifteen minutes soon enough?"

"Right. And wear long pants. The bugs may be hungry."

Twenty minutes later they were driving south toward Everglades National Park. Nikki had dressed in blue jeans, a yellow halter top, and a print blouse. She looked like a teenager. Michael, in slacks and a sports shirt, could easily have been taken for her uncle or big brother.

"Why did you think I wouldn't want to go?" he asked her. "Aside from the fact that it's obviously a major inconvenience."

Nikki cursed the fact that she blushed so easily. She knew she owed him an apology for last night. "Michael—" she said uneasily. "I wanted to explain—"

"Yes?" He cocked an arrogant eyebrow at her.

"Well, I'm sorry. About last night—I don't know what got into me."

"Don't you?" he interrupted.

"You're not making it very easy, are you?" Nikki said heatedly. "I know you have a right to be angry—"

"Who said I was angry?"

She hesitated. "You were so grouchy this morning."

"I'm always bad-tempered in the morning. You wanted to pay me back and to a certain extent you succeeded. But you're still in the minor leagues, Nikki," he gibed. "You have a long way to go before you're ready for the big time."

"Aren't *we* pompous this morning!" Nikki retorted. But she was relieved by his forgiving attitude, and her voice was all Florida sunshine as she added, "I refuse to let you goad me today, Michael. Frankly, you don't seem exactly in tune with Mother Nature. I pictured you as a child who played ball on the dirty sidewalks of New York. The only animals you would have seen there are pigeons."

"You're the one who grew up in New York City," Michael said. "When I was a baby my parents moved to a house in the suburbs. A few years later they bought a modest little twenty acre estate in Bronxville. I broke my nose when I fell out of a spruce tree. I was always bringing home garter snakes, turtles, and anything else that moved."

"I can't imagine it," Nikki said tartly. "I thought the only moving creatures that interested you were female and over twenty-one."

He grinned. "What I like about you, Nikki, is that you're so submissive, so eager to please me."

His remark had been made in jest, but Nikki realized with a start that at least part of it, the second part, was true. She did want to please him this morning. It had become important to her to be on good terms with him. She very much wanted him to like her for herself, not to indulge her because she was part of some incredible business arrangement and needed to be charmed into docile acquiescence, or because she provided a temporary and amusing diversion from his hectic schedule.

She looked straight ahead at the road; she wanted to explain her feelings but lacked the courage to turn toward him.

"The funny thing is, Michael, that I really don't want to fight with you. If—if you asked me to do some favor for you, I gladly would. You can be very nice when you want to be. And last night—well, frankly, you could have—but you didn't. My mother thinks the world of you, too. I don't believe you're serious about having a child. But you—you amuse yourself with me. I want to be taken seriously. I—I don't want to go to bed with someone who thinks—who thinks I'm a plaything." Her voice had dropped to a near-whisper.

If Nikki had hoped that this husky appeal would meet with a positive response, she was mistaken. Michael said nothing. Instead he reached over and switched on the radio, fumbling with the dial until he found a raucous rock music station.

They entered the Park and drove to the southern tip. Nikki was tense with self-reproach as they went into the Visitors' Center. Everything had been going so well; why hadn't she kept her big mouth shut! Now Michael was angry and she really didn't understand why; she only knew that it was somehow her fault.

They walked around looking at the displays. Here was the last sub-tropical wilderness in the United States. Nearly one and a half million acres of water and land, mostly swampland, it was noted for its often rare animal and plant life. Nikki looked at the pictures, but her mind wasn't on them.

Michael bought tickets for a guided boat tour through some of the park's many waterways. As the guide pointed out different types of trees and flowers, Nikki began to become interested in the trip. Everyone was trying to spot birds and animals. When Nikki noticed a great white heron before any of the others, she excitedly turned Michael's head toward the four foot long bird.

By the time they stepped off the boat, much of the earlier tension had dissolved. They ordered sandwiches in the park restaurant, for the most part eating in silence. Nikki was afraid to say anything; she could not face a resumption of hostilities.

After lunch they strolled leisurely along some of the

trails, trying to see who could spot more wildlife.
Anhingas, or water turkeys, abounded. They watched
one glossy black and silver bird as he perched on a
limb, soared out of sight and then returned several
times, finally deciding to splash into the water. Alliga-
tors, sea turtles and fish were common. And they were
lucky enough to see one of the park's crocodiles; the
reptiles had become quite rare, and had been designat-
ed a protected species by Congress.

Michael again switched on the radio as they drove
back to Key Biscayne, but this time he tuned in a
classical music station. To the strains of Beethoven's
Pastoral Symphony, Nikki fell asleep. She woke up, her
head on Michael's shoulder, as they drew to a halt in
the Craguns' driveway. With a sleepy smile at Michael,
she moved away and got out of the car.

Ann and Jake Cragun were sitting by their pool; they
suggested that Michael and Nikki cool off with a short
swim before dinner.

They changed into their suits and Michael swam
several brisk laps, then called out to Nikki to join him
with the customary, "Come on in, the water's fine."
Too comfortable to budge an inch, she declined.

"Suit yourself," he shrugged, polite but distant. But
a few minutes later he emerged from the water, and
before Nikki realized his intention he had scooped her
up and tossed her into the pool. He dove gracefully in
after her; she retaliated by splashing him in the face.

"If you get into a water fight with me, Miss Warren,
you're going to lose. Why don't we let you work out
your hostility with a game of volleyball, hmm?"

She nodded, and Michael set up the net across the
shallow end of the pool. The bottom was sloped and
Michael stood in the deeper water. There followed a
free-for-all of splashing, laughing, hitting and diving for

the ball. Nikki's hair came undone; she complained that it kept getting into her eyes and blamed that for ruining her game. But in fact Michael's height and reach were too great a physical advantage; he spiked the ball viciously. Obviously he played to win.

Much to Nikki's disgust, she managed to win only a few points. She knew it was childish of her, but once the game was over she felt furious about being beaten so badly. When Michael put his arm around her shoulder, she shrugged it away angrily.

"I can't believe it!" he yelped. "You're actually sore about losing."

Her emotions heightened, she was suddenly aware of how handsome he was, standing next to her wearing only a body-hugging blue swimsuit and a towel draped around his shoulders.

"Go away. You didn't play fairly. You could have let me win a few more points!"

"Now if that isn't some sort of perverse feminine logic. You didn't really want me to *let* you win?"

"Well, I would have," Nikki insisted, "if you hadn't told me you were doing it!"

"Women!" It was an exclamation of disgust. "What you need is a chance to cool off." And he picked her up, easily overcoming her squirming efforts to escape, and laughingly tossed her back into the water.

Michael's parents, relaxing in their lounges, had witnessed the entire exchange. Under the circumstances, Nikki summoned up as much dignity as she could and climbed tremblingly out of the pool. Her gaze was stony as she took a towel from Ann and allowed herself to be led into the house.

"I know I should keep my mouth shut, Nikki," Ann said, "but quite frankly, dear, I can see you're in love with Michael. And he may not know it, but he feels the same way about you. Give him time—he'll come around." She smiled brilliantly. "And when he does, Jake and I will be pleased to welcome you into the family. Now run along and get out of that wet suit."

Nikki mumbled some suitable words of thanks and escaped to her room. Ann Cragun's kindness had certainly taken her mind off her anger at Michael's tactics, but it had been replaced by something worse. In love with Michael Cragun? Impossible, she thought. He was arrogant, domineering, conceited, and selfish. He didn't know how to be tender or sweet unless he wanted something from someone. There was no point in denying the strong physical attraction she felt for him, but that wasn't the same thing as love. She had not even known the man long enough to fall in love with him.

Once they were back in New York she would have nothing more to do with him. Her sense of self-preservation dictated that. She dismissed the threats he had made; she had learned this weekend that he was a perfectly rational person; he would never actually carry out such outrageous blackmail.

At the airport, Jake and Ann warmly hugged their son good-bye, telling him not to wait too long until the next visit. To Nikki's embarrassment, both of them kissed her as well, saying they would look forward to seeing her again.

As she sat on the plane with Michael, Nikki considered the weekend she had just spent. It had ended as it had begun—with a long, confidential business conversation. In line with the instructions of the lady of the house, the men had acted as chefs. Nikki was amazed when Michael appeared with a chocolate soufflé he had made for dessert, although he solemnly claimed that it was absolutely the only dish he knew how to prepare.

Dinner time discussion centered on the economic picture and long-range planning for CAI.

Nikki had been forced to acknowledge that Michael had some positive qualities. He had listened to his parents' ideas, and even hers, with an open, receptive mind. He was extremely intelligent, a more than competent business executive who Nikki sensed had won the loyalty of the CAI staff. He enjoyed a warm relationship with his parents, although he insisted they stay out of his personal life. He had a sardonic sense of humor, even if Nikki had smarted under the lash of his sarcastic tongue on a few occasions.

She remembered that on Friday night she had resolved that she would learn more about him. She did not feel that she had been permitted more than a glimpse under the layers of defenses, although why he should be so guarded she could not fathom. He had come from a loving home; he had money, the best education, every possible opportunity, and devastating good looks. His job interested him, yet he often seemed bored with life. He could have had his choice of almost any woman he wanted, but that seemed to bore him too.

Nikki finally concluded regretfully that an analysis of Michael Cragun's character was quite beyond her abilities as a psychologist, and absent-mindedly accepted a magazine from the flight attendant.

Michael sat beside her, wading through a report that had "confidential" stamped all over the front cover in red ink. He seemed to be totally absorbed in his work. Although Michael had said he would discuss the future with her tonight, he seemed to have no intention of doing so.

After a perfect flight, they were met by a waiting Henry. They had been driving for a little while when Nikki realized that they were taking a route that led west toward Manhattan and not north to the Bronx, where her mother's apartment was located.

"Where are we going?" Nikki asked nervously.

"Home."

"This doesn't go to my apartment." A tremor of apprehension shivered through her.

"We're not going to your apartment. You're coming home with me." Michael sounded matter-of-fact, even bored.

"No, I'm not," she objected. "If you mean to have that talk you promised, forget it. It's late and I'm tired and I want to go home."

Now Michael turned to look directly at Nikki, and she could see the uncompromisingly steely expression in his eyes. His tone matched. "What I mean, Nikki, is that you're going to live with me. I decided that before we left, and the matter is settled. There's no reason why Henry should have to drive all the way to the Bronx to pick you up every day."

"So tell him not to!" Nikki all but screamed. "You can't dictate to me. I've taken the subway for years and survived. I don't need your chauffeur."

"Interrupt me again, and I promise you you're going to regret it," Michael said evenly. "I want you to understand your situation. The furniture has been moved out of your mother's apartment and stored. Your personal possessions have been taken to my place. Your landlord has agreed to cancel the rest of the time on the lease as soon as he re-rents, which he probably already has because there's a waiting list for that building. Does that clear things up for you?"

Nikki dug the nails of one hand into the palm of the other, and tried to calm herself. Henry Merola must have made all these arrangements while they were in Florida. It was pointless to appeal to him for help. And losing her temper had never gotten her anywhere with Michael.

"Michael," she said falteringly, "why are you doing this?"

"Maybe I can't do without your company," he drawled.

"I don't think that's very funny. You could at least explain to me," she all but begged.

He refused to take her resistance seriously, saying ingenuously, "You know how many women would like to be in your shoes? Look at it this way. It's more convenient. You won't have to do any cooking or cleaning. Hell, I'm doing you a favor, Nikki. You should thank me."

"I suppose you've decided that I should reward your generosity," Nikki responded acidly.

"Nikki," Michael said in an icy voice, "I enjoy making love to women who know how to please me and are anxious to prove it to me. Will you tell me why you persist in thinking that I crave the body of an inexperienced, repressed little virgin?"

"Because of your damn son or daughter, that's why!" she spat out at him, hurt by his description of her.

"I've told you more than once that you're in no shape for that. I'm no saint, but I'm hardly a sex starved adolescent either. Nonetheless, next time you throw yourself at me, I'm going to accept the invitation." He took in her livid expression, then chuckled rakishly, "Let me hasten to assure you that Henry and Suzanna live with me. They'll protect your precious virtue."

Nikki said nothing. She wavered between wild fury at the man's sheer gall and acute humiliation at his assessment of her. But she knew there was no way she would live in the same apartment with him; he could hardly tie her up and keep her there. For tonight, it seemed she had no choice, but tomorrow she would leave. And, she told him silently, there isn't anything you can do to stop me!

Nikki was hardly surprised to find that Michael Cragun lived in one of the most exclusive areas of Manhattan. He owned a duplex penthouse apartment on Fifth Avenue, facing Central Park. It was an older building which had been converted into a cooperative. Obviously a healthy sum of money had been spent on renovating the two-story apartment. It occupied the entire northwest corner of the building. Nikki had to admit that it would be pleasant to get out of bed in the morning and look out over trees and flowers and water instead of dirty concrete and brick.

Michael's home, she grudgingly acknowledged to herself, was beautiful. He had half-dragged her around the downstairs, not caring that she was angry and unresponsive. A large, tiled entrance hall divided the apartment between a kitchen and dining room on one side and a living room and library/study on the other. The Merolas lived in a small apartment off the kitchen. A staircase led upstairs to a huge master bedroom suite and two guest rooms.

In the living room, Michael's taste for oriental carpets and antiques was in evidence, but this was combined with comfortable modern furniture. The color scheme of blues, natural tones and off-white would have been very soothing if Nikki had been in the mood to be soothed. Upstairs, Michael did not include his own room on the tour but she could judge the size of it by the fact that it was the only door on that side of the hall.

Michael ushered her into "her" room, saying drily, "I'm sure you'll feel right at home, Nikki." The reason for this statement was clear: all her own things—family pictures, a tattered yellow teddy bear she had received for a sweet sixteen present, her books, clothing, gifts she cherished—were in the room.

Nikki's only concern at the moment was to be left alone. "Thank you so much. You're so very thoughtful. Now kindly get out of *my* room," she smiled her most insincere smile, "please."

Michael grinned at her. "With pleasure, ma'am." Whistling cheerfully, he left the room.

Alone, Nikki inspected the bedroom. The furniture was simple and caned and quite to her taste. Soft apricot drapes blended with sand-colored carpeting; muted floral spreads covered the twin beds. A sliding glass door led out to a terrace which faced the park. New York City was often sooty and polluted, but on a clear spring day it would be heavenly to sit outside and people-watch. A treacherous thought entered her head. How easy it would be to stop fighting, to live here, to agree to whatever Michael wanted of her.

He seemed to be able to work things out to suit himself. She had scarcely bothered to wonder how he managed to arrange with her landlord and the building superintendent to terminate her mother's lease and move everything out of the apartment. Perhaps it was useless to make plans to leave, he would find a way to sabotage them.

Becoming Michael Cragun's mistress, even bearing his child, was hardly the worst fate in the world. But, Nikki asked herself, for how long? Nine months? A year? By then she might be desperately in love with him, only to be traded in like last year's Cadillac.

She lay awake for hours, planning her escape and rehearsing what she would say to him when he confronted her.

Chapter 5

Nikki Warren had been a composed little girl who had grown into a self-contained young woman. But feigning cheerful resignation to her fate called for more control than she had ever imagined. Fortunately her morning confrontation with Michael was brief. He left for work at 7:15, just as she was getting up. They passed in the upstairs hall; she received only a grunted acknowledgment of her bright "Good morning."

An hour later Henry drove Nikki to work; of Mrs. Merola she saw nothing, and she wondered if Henry's wife were a figment of Michael Cragun's resourceful imagination. And she would not appeal to Henry for help; indeed, she took care not to arouse his suspicions by attempting this maiden jailbreak without even so much as an extra shopping bag. She left with the clothes on her back and nothing else.

Nikki had come to the distasteful conclusion that she could not possibly succeed in disentangling herself from Michael Cragun without some help. She hated to involve anyone else in her problems, but she had no choice. Erika had immediately commented on Nikki's sunburned face Monday morning. "Where did you spend the weekend?" she asked. "Under a sunlamp?"

"Erika, I've gotten myself into a terrible mess," Nikki burst out, abandoning her intention to calmly explain her position to her friend. "I don't know where to turn." To her dismay, she was almost in tears. It was much easier to think about her situation than to talk about it.

An alarmed Erika got up from her desk and rushed over to lean against Nikki's. "My God, Nik, what's the matter? You're not pregnant, are you?"

Nikki began to laugh almost hysterically at Erika's unwittingly ironic question. "Oh, Erika, I can't talk about it here. Can I go home with you tonight?"

"Of course," Erika told her soothingly. "You can stay as long as you want. There's an extra bed in my room; Nancy, my roommate, won't mind."

For the rest of the day, Nikki threw herself into her work, hoping to push Michael Cragun from her mind. It was impossible. Only when she and Erika were settled on the train to Yonkers did Nikki begin to explain the events of the last two weeks. There was no point in enlisting Erika's aid unless Nikki told her the truth. Erika had to understand that she was not suffering from some paranoid delusion but that she had reason to feel desperate. So she omitted nothing, even though her face flamed as she recounted the details of her encounter with Michael in Jeremy Whitney's private hallway. The list of adjectives Michael had used to describe her was even worse to repeat.

By the time Nikki finished relating the tale of her virtual kidnapping the night before, her friend was transfixed. By now they were in Erika's car, riding back to New York to visit Nikki's mother in the hospital.

"My God, Nikki, it's like a soap opera. If I read it in a book, I wouldn't believe it. But it doesn't make sense. Granted you're a beautiful girl, but to a guy like Cragun that's nothing unusual. Why is he going to so much trouble to make you do what he wants?"

"You have to realize it isn't any trouble for him,

Erika. All he has to do is make a phone call or give Henry an order, and poof—his wish is granted." Nikki's sense of humor had begun to return; she felt incredibly relieved to be sharing her troubles with someone.

"I don't buy that," Erika frowned. "He's a busy executive; it's got to be taking up his valuable time. He must have a reason for bothering."

"Okay. So maybe it amuses him. I told you I thought it was all a big game to him."

Erika screwed up her face in thought. "I can accept that. At least I can accept that in the beginning he might have gotten a kick out of scaring the stuffing out of you. You really are an innocent, Nikki. But why go on with it?"

"*You* ask him," Nikki suggested. "I certainly can't get a straight answer."

"Look, Nikki, his mother told you she thinks he's in love with you but he's a bit slow about figuring that out. He's been her son for thirty-two years; what makes you think she's not right?"

"Oh, Erika!" Nikki moaned. "She thinks I'm in love with *him*, and that's—ridiculous. A crush—I admit to a crush. But no more than that. Besides, I told you some of the things he said—that I'm inexperienced and repressed. He never even tried to touch me last weekend and God knows I encouraged him. It's all a lark or a dare or something. There's no point in trying to guess his motivations. I just want to stay with you for a while. And—and if you would come with me to get my things on Saturday—I can't face him alone."

Erika gave Nikki's arm a reassuring squeeze. "Don't worry. You can even borrow some of Nancy's clothing in the meantime, you're about the same size."

The girls found the usually cheerful Mrs. Warren

looking especially pleased about something. She extended a warm greeting upon being introduced to Erika, then promptly informed her daughter, "You just missed Michael. How was Florida?"

Nikki blanched. Thank goodness it had taken them a long time to park. The thought of running into Michael so upset her that she found herself shaking; she tried to cover up by adopting an uncharacteristically chatty manner. She prattled on about the beautiful weather, the Craguns' home, the dinner party Saturday night, her conversation with the Newhouses, her trip to the Everglades. Mrs. Warren clearly mistook her nervousness for wild enthusiasm.

"Michael is such a sweet boy," Pam Warren said. It was on the tip of Nikki's tongue to point out that Michael was hardly a boy. But she held her tongue. Her mother continued the litany of praise: "He takes good care of you. You're very lucky. His apartment—it's nice?"

Nikki felt her face heat up. She stared down at Pam Warren, wondering how much her mother knew. When she spoke, her voice was vaguely strangled. "Mom, did he say anything to you about *your* apartment?"

Her mother nodded, a look of satisfaction on her face. "Of course. We talked Friday. He suggested giving it up. Very sensible, I thought."

Nikki silently cursed. The man always seemed to be one step ahead of her. He knew perfectly well she would never risk upsetting her mother by explaining their real relationship. But she was astonished that her strait-laced mother would accept so calmly the idea of her living with Michael Cragun. Even if Mrs. Warren did consider him a candidate for sainthood.

She phrased her inquiry carefully. "Mom, you *do* realize I've sublet my apartment until June?"

Mrs. Warren nodded, her face a study in unconcern.

"You think it's proper for me to live with Michael?" Nikki asked, her voice slightly hysterical.

"A different sort of man—I'd disapprove. But Michael is so honorable. He would never take advantage." She smiled tiredly.

Nikki did not pursue the matter. She would tell her mother that she was staying with Erika next time she visited. She bent down to kiss Pam Warren good-night and watched as her mother contentedly closed her eyes and fell asleep.

Tuesday found Nikki and Erika hard at work. Every time the phone rang, Nikki's heart began to thud erratically. But there was no call from Michael Cragun. Fortunately, she was too occupied by the pressure of work to sit and feel sorry for herself. Nonetheless, when they left for the day, she was physically sick with the fear that she would find him waiting for her, ready to punish her for her rebelliousness in some unspeakably diabolical manner. Erika hovered protectively close, but there was no sign of him. Nikki glanced up the street. The Cadillac limousine was not in the usual spot. Yesterday she had been relieved by that. Today she had the uncomfortable suspicion that Michael had expected her to leave, and was prepared for it.

She called her mother when they got to Erika's and lied, saying that she had had a terribly hard day at work and was too tired to visit. Pam Warren was sympathetic but disappointed. Nikki, plagued by guilt, realized she could not avoid her mother out of fear of running into Michael. She assured Mrs. Warren that she would visit the next evening.

When Wednesday morning passed with no word from Michael, Nikki began to relax. She made herself believe that he had gotten tired of playing with her and would leave her alone.

She stepped out of her office before lunch and returned cheerfully, only to be greeted by a peculiar look from one of her co-workers. "Well, you sure move in fancy circles," the woman told her. "You've got a visitor in your office."

Nikki didn't even ask who. She knew. It was cowardly, but she wanted nothing so much as to run out of the building and never return. Only the fact that Erika would be there to lend her support gave her the courage to go into her office. Trembling and nauseated, she opened the door—and surprisingly came face to face with Peter Delavan, the publisher, who stood in the center of the room—alone.

He looked at her, his expression defensive. "I've come to take you out to lunch. No—don't ask questions yet; just come."

Nikki, shaking, grabbed her purse and a sweater and followed him out. She could feel everyone's eyes on them.

Spring had put in an unexpected appearance in New York. Over the past few days the snow had melted; today the temperature had climbed to a balmy 60°. Peter Delavan, obviously embarrassed by what was to come, tried to make small talk about the beautiful weather. Nikki was completely unresponsive. She knew her silence annoyed the publisher, but she was too upset to care.

Not until they were seated in an intimate, dimly lit Italian restaurant did Peter Delavan begin to talk. "I've known Michael since we were in business school together, Nikki. We became friends—very good friends. He's done me some special favors over the years. Frankly, our business reporting is much stronger because of him." The publisher sighed and slowly shook his head. "Now Mike is calling in the I.O.U.'s. He says to tell you that he's giving you until Friday. I don't know exactly what that's supposed to mean, but unless he tells me otherwise, on Monday you'll be back in the advertising department. You know you're still on . probationary status in personnel. The step after that is to fire you—not easy because of the union, but Michael has a way of getting what Michael wants."

He took in her stricken face. "Nikki, I wish you would tell me what this is all about."

She silently shook her head. Even had she been inclined to confide in him, no words would have come out.

"God, I feel like a beast," he groaned. "But I have no choice. I owe him too much. I suppose you two had some kind of a fight. Michael has no right to use me to pressure you. But he backed me against a wall. Nikki, you must realize that he has to love you very much to go this far."

At that statement she came back to life. "You're wrong," she said bitterly, "he doesn't love me at all."

"Come on. I saw you two together. I've known Mike for ten years and I've never seen him so possessive—"

Nikki stood up abruptly, interrupting Peter Delavan with a mumbled, "Excuse me. I'm not hungry." Without waiting to see if he would reply, she hurried out of the restaurant.

Standing on the busy sidewalk, she took several deep breaths and gave herself a pep talk. Hadn't she been expecting something like this? She was a fool to let it disturb her. If they tried to fire her, she would fight it. Instead of making her cave in like a spineless coward, this incident should be stiffening her determination to fight. Michael Cragun had to be totally unprincipled to force a close friend to do his dirty work. She held no grudges against Peter Delavan, the man was apparently convinced that Michael was desperately in love with her. He probably assumed he was being used to patch up a lovers' quarrel.

She forced herself to consider her options rationally. It made sense to proceed as if Michael's threat were just so much hot air, as if he would never go so far as to have her demoted or fired. He had expected her to leave and was trying to bluff her into coming back. After all, she had seen for herself how competitive he was. It would be out of character for him to concede the game without a fight. So she would fight! With new determination, she returned to her office.

Michael was not at the hospital that night; Nikki was actually disappointed because by then she was itching for a confrontation. Her initial despair had turned to hard fury.

Thursday afternoon brought a call from the young man whom Michael had put to work managing Mrs. Warren's store. He cleared his throat before he began to speak; his cough sounded defensive, nervous.

"Uh—Miss Warren. This is Bob Jacobs. Have you spoken to Mr. Cragun lately?"

Nikki knew what was coming. She resigned herself to it. "No, I haven't."

"Well, I'm really sorry, but he insists that he needs me back at headquarters. I can stay through tomorrow afternoon, but after that—" His tone was abjectly apologetic.

He was so genuinely contrite that Nikki felt sorry for him. "Of course, I understand," she said sweetly. "Thank you for all your help. I've—I've appreciated it enormously." She permitted herself a small sigh before resuming her work.

When the hospital business office called at 4:15, it was hardly a shock. A Mrs. Cardin explained that Mr. Cragun had called to complain about an error concerning her mother's bill. "He was terribly angry and demanded to know how his check, which was a donation, got applied toward your mother's account. He all but accused you of dishonesty in the matter, Miss Warren. He asked how much you owed, and when I told him over $5000 he said your mother should be moved to the public ward of our sister hospital. I'm terribly sorry, but the Craguns are influential at this hospital; I'm going to have to have the bill settled immediately."

"But, Mrs. Cardin," Nikki protested, "none of that's true. You must realize that. I can't explain any more than that. But I *do* promise I'll pay back every cent eventually. Can't you wait?"

"I'm afraid not. Frankly, it doesn't matter what the truth is. We can't afford to alienate a family that's given us millions of dollars. It's not as though your mother would have no care at all. But I have to tell you in all honesty—she would be better off staying where she is if you can possibly manage it."

Nikki swallowed back the tears which threatened to spill over. "I—I understand. I need some time to think. Can I—can I call you back tomorrow morning?"

This proved agreeable to Mrs. Cardin, who told Nikki she would expect her call at 9:00 A.M.

After each new setback, Erika Berger had provided encouragement. But as they left work Thursday afternoon, there was nothing she could say or do to improve an impossible situation.

"It all comes down to this, Erika," Nikki said with a forlorn shake of her head. "Either I do what's best for my mother, or I think first of myself."

"But it's moral blackmail!" burst out Erika, incensed. "Couldn't you go to the police, or get a lawyer?"

Nikki screwed up her mouth. "Do you think anyone would believe me? And even if I could afford a lawyer, do you know how long court cases take? It would be months. I have to decide *now*."

She glanced up the street. "I don't think I have any choice, Erika. In a way it's funny. When mom had a relapse, I felt so powerless, I hated not being able to control everything. But compared to how I feel now, I was practically a free spirit then. Listen, I want to thank you for all you've done."

"You mean you're going to let him boss you around? Go back there as if nothing had happened?"

Nikki nodded.

"When?" her friend demanded to know. "Tomorrow?"

Nikki pointed up the street to the black Cadillac sitting two blocks away. "Do you see that limousine up there? It belongs to Michael. That's his chauffeur driving it. For the past three days I've looked up the street and he hasn't been there. Today he is. Just as if Michael knew when I'd be ready to give up." She gave a resigned shrug. "Why put off until tomorrow what I can do today?"

Erika hugged her. "If there's anything I can do—"

Nikki shook her head. Then she walked slowly up Eighth Avenue to the black car, opened the door without a word, and got in. Henry Merola started the engine.

They drove straight back to Michael Cragun's apartment. Nikki said nothing, asked no questions, just climbed up the stairs to her room. For a few minutes she cried soundlessly, then pulled herself together. She wasn't going to let *him* see her red-eyed and distraught.

An hour later there was a gentle knock on the door, which was opened before Nikki could object. A smiling, slightly plump woman entered carrying a tray

with dinner on it. "I'm Suzanna Merola," she introduced herself. "The boss will be home late tonight, so I
thought I'd bring you a little something to eat." Her
manner was warm, motherly. "Now, Miss Warren, I've
raised two girls of my own. You eat. You'll feel better, I
know."

Nikki doubted it, but Suzanna seemed so concerned
for her that she said politely, "Thank you, Mrs.
Merola. I'll try."

The minestrone soup smelled delicious, and as Nikki
sipped it she suddenly felt ravenous. She would not
have to see Michael tonight. Time enough to mope
later, she told herself as she attacked the veal scaloppi-
ni alla Marsala, fresh asparagus, and cherry cheese-
cake. She avoided dwelling on what would happen in
the morning. For a time she listened for Michael's
footsteps on the stairs, but heard nothing. Eventually
she clicked on the color television set in her room and
tuned in an old western; within fifteen minutes she had
fallen into a troubled sleep.

She woke up early, at 5:30. She was still in her
clothes, which felt uncomfortably sweaty, but the
television set had been turned off and someone had
covered her with a light blanket. The tray was gone
from the night table; Suzanna Merola must have looked
in on her.

Nikki got out of bed and peeked into the hall.
Michael's door was closed; she heard no sound except
the relentless cacophony of the city traffic outside. She
quickly slipped into the bathroom to shower away the
grimy feeling from her body. The shampoo and cream
rinse she used sat on a corner of the tub; Michael
certainly thought of everything. Nikki's sense of humor
was too well developed not to be amused by that; with a
shrug she began to wash her hair.

Half an hour later she carefully emerged from the

bathroom, dressed in a floral print shirtwaist, and went downstairs in search of a glass of juice.

To her dismay, a bathrobe-clad Michael was sitting at the kitchen counter, drinking coffee and reading the morning paper. He glanced up as she came in, then stuck his nose back into the sports section of the New York *Sun*. A moment later, as Nikki stood rigid, willing herself not to flee, he tossed the paper onto the blue-tiled counter.

"Good morning. Sleep well?" He could have been talking to the neighbor's dog for all the interest his tone held.

Nikki had been unsure of how she should react when she saw him again. She refused to lose her temper; every time they had quarreled, he had won. She settled on frigid politeness. She would not give him the satisfaction of knowing that deep inside she was quaking. Her palms were moist, her heart pounded so rapidly that she felt faint, and worst of all, in spite of the way he had treated her, she was still intensely aware of his pure male sex appeal. No man had ever affected her this way before; why did it have to be him?

She strove to match his indifference in her reply. Perhaps if he were convinced that his game bored her, he would leave her alone.

"Yes, thank you." She walked to the refrigerator and took out a bottle of grapefruit juice.

"Miss me?" he went on. He was following her with his eyes as she filled a glass with juice.

"No." His arrogance was so astounding that she had to restrain the impulse to throw the juice in his face.

He flashed her a lazy smile, and said softly, "Angry with me?"

Her tone was airy. "You're not worth expending my emotions on."

"I'm not? What a blow to my ego!" he taunted.

Nikki ignored the sally, finished her juice, and placed the empty glass in the dishwasher. As she straightened up her eyes met Michael's.

"You owe me a thank you," he drawled. "I covered you last night. I would have undressed you," he grinned, "but I didn't want to risk being crazed by unquenchable lust."

She swallowed the angry retort which had leapt to her lips and lifted her shoulders. "Sure. Thanks." She turned to leave the kitchen.

"Can't bait you this morning, eh, Nikki?" His laughter rang out from behind her. "Okay, come back here and sit down. I'll make you some breakfast."

"I thought you only knew how to make a chocolate soufflé," she snapped out sarcastically. The words had slipped out unbidden, and she silently reproached herself for letting him goad her.

"Chocolate soufflé and scrambled eggs. Also toast and coffee." She heard the stool scrape against the floor as he got up, and the next moment felt his hands on her shoulders.

"How about it?" he said softly, his tone more appropriate to an attempted seduction than an offer to cook her a meal.

She shrugged away his hands. "No thank you. Seeing you again has made me lose my appetite." He made no attempt to stop her from leaving the kitchen.

Alone in her room, Nikki relived the scene that they had just played. Michael had been alternately cool and mockingly amused. If only he had gloated over his victory, pure fury might have coursed through her. As it was, she found her emotions a mixture of anger—and something else. Her attempts to hand Michael a cool setdown were a source of casual entertainment to him. It did not seem to matter to him that his actions might upset or even frighten her. Nikki was used to being important to people and it was disconcerting to be little more than the mouse to Michael Cragun's cat. She impatiently wiped away the wetness from her eyes with the back of her hand and turned on the television.

Fifteen minutes later as she sat cuddled up in a comfortable armchair watching the morning news, Michael pushed the door open. Nikki continued to stare at the screen.

When Michael silently approached her chair and reached down to turn her face up to his, she was so startled that her whole body tensed. "Calm down. I won't attack you," he scowled. "Judging from the warmth of your welcome downstairs, you'll be glad to hear that I'm going away for the weekend. I came in to say good-bye. I'm leaving directly from the office, so I won't darken your door until Sunday night."

"It's *your* door," Nikki answered in a low voice. "You can do what you want with it."

"Don't you want to know where I'm going?" he asked teasingly.

"Mr. Cragun," Nikki said hoarsely, "I existed very happily for twenty-three years without being aware of you or your movements." To her dismay a tear slipped out of one misty green eye. She sniffed noisily and angrily brushed it away. "Oh, just leave me alone," she snapped.

Michael paid no attention to her angry command, but sat down on the arm of her chair. With a long, tapered finger he carefully brushed away a second tear that was wending its way down her cheek. When she jerked her head away and began to get up, he put a firm hand on her shoulder to prevent her.

He said huskily, "Don't cry, Nikki." His hand moved from her shoulder to stroke her hair, silky and fragrant with the scent of the herbal shampoo she had just used.

Michael gazed at her face, and continued to play with

her hair, twisting it around his fingers. Nikki sat, her face expressionless, paralyzed by indecision. His casual touch had aroused her deeply. The warmth had started in her face and traveled down to fill her whole body. Michael watched for the response he must know she was aching to give. A picture flashed through her mind: she was reaching up, pulling his mouth down to hers.

But sanity reasserted itself as a second picture was recalled: she was in Peter Delavan's hallway, and Michael Cragun was kissing away her anger. He enjoyed manipulating her emotions; it would suit his convenience to make love to her, then leave for the weekend certain that she would slavishly await his return.

Abruptly she flung herself out of the chair and stood with her back toward him. She heard him mutter a low curse.

"You know damn well you like it when I touch you. I'm not some kind of monster. So what the hell is the matter?" he demanded impatiently.

Nikki faced him, and raised her chin several degrees higher. "Get out of here." She was trembling, both with hurt and self-disgust. "I despise you. You forced me to come back here and I hate you for it. You're— you're wrong. I can't stand it when you touch me. I'd—I'd have an abortion before I'd bear any child of yours. Does that answer your question?"

If it did, Michael was not about to say so. He stalked out, slamming the door behind him.

Nikki arrived at her office an hour and a half later, outwardly calm. She had been horrified by her last angry words to Michael. Abortion was not something she took lightly, especially not for Michael's child— hers and Michael's. She didn't hate Michael. She was far from being repelled by him. Her feelings were confused, jumbled. She was grateful that he would be gone for the next three days—three blessed days of freedom.

Erika soon joined Nikki, and her face communicated her deep concern, concern that Nikki appreciated. But Erika's anxious questions evoked no desire for confession on Nikki's part. She was reassuring but noncommittal, intent on avoiding an instant replay of the morning's events.

Later in the day a messenger brought Nikki an envelope with CAI embossed in the upper left corner. She slit it open to find a brief note and a wad of bills.

"Nikki—" it ran, "hope you enjoy your weekend as much as I intend to enjoy mine. Enclosed is some pocket money. M. P.S. You can thank me properly when you see me."

Angrily, she counted the money. $200! Who did he think he was? And *what* did he think she was? She resolved not to spend a penny of his money. She'd spend the weekend reading and visiting her mother. Michael could just keep his bribe—or whatever it was! She flushed as she stuffed the money in her purse, hoping that no one had seen the bills or the letter.

Her Friday visit with her mother was a short one; it had been interrupted by a doctor who introduced himself as Stephen Rowland. Nikki, feeling that she was intruding, had excused herself and promised to return later that weekend.

She spent the weekend window-shopping and reading, and returned to the hospital on Sunday afternoon. Dr. Rowland, a balding, friendly man who appeared to be in his middle fifties, sat by Pam Warren's bedside, holding her hand. He turned and rose as Nikki entered. She bent down to kiss her mother hello. Then she looked at Dr. Rowland, wondering who he was and what he was doing holding her mother's hand.

He smiled at Nikki. "I met your mother last winter when I dropped by her store to pick up a new electric frying pan, Nikki. I—uh—had occasion to shop there from time to time after that. Especially after I found out via the hospital grapevine that she was a widow. You know, I became a surgeon because I've always liked a captive audience. And for the next few weeks, I've got one."

To Nikki's surprise, he took her by the arm and led her into the hallway ,where he told her in a sincere, low tone: "I thought your mother was a very special lady when I got to know her a bit in the store. Her stroke hasn't changed my opinion—only re-enforced it." He studied Nikki's face thoughtfully, as if waiting for some sign of approval.

Nikki certainly shared his high opinion of her mother, but she felt terribly embarrassed to be treated like Mrs. Warren's mother by a man old enough to be her father.

When she made no reply, Stephen Rowland continued, "You know, Nikki, now I know how my son-in-law felt when he asked my permission to marry my daughter. But I just want you to know that I'm reasonably respectable. My wife died two years ago. I have a twenty-five-year-old son in addition to my daughter, and my intentions are thoroughly honorable."

Nikki's face had warmed to such an extent that she felt it must match her lipstick. She blurted out, "Are you asking my permission to propose to my mother?"

"No, no," he said quickly. "We don't know each other well enough for that. But I've heard from some of the nurses that you're protective of your mother. I don't want you to worry that I'd hurt her in some way."

Nikki unclasped the hands that had been held

clenched in front of her and shyly looked up at Stephen
Rowland. "You have an excellent bedside manner,
doctor." She extended her right hand toward him. "It's
been nice to talk to you. I have a feeling we'll be seeing
a lot of each other."

He took her hand in both of his, winked, "You can
count on it," and, with a wave at Pam Warren, hurried
down the corridor.

Nikki returned to the room and sat down by her
mother's bedside. Her expression was thoughtful. "I've
got something to say, Mom." She paused to organize
her forthcoming lecture and found her heart was
thumping erratically. All her life she had relied on her
mother's advice; this role reversal made her uneasy.

"That man likes you, mom. He's a doctor, so he
knows the score on your health. No, don't interrupt
yet, because I want to finish first. I know you and dad
were unusually close. The pain of losing him—" Her
voice trailed away, anguished, but a moment later she
continued firmly, "But you're only forty-five. Promise
me you'll keep an open mind."

Pam Warren regarded Nikki tenderly, the dewiness
of her daughter's eyes reflected in her own. "Some
speech. You know, I was worried about *you*. You and
daddy—were—special too. I've always believed life is
to live, not run from. I will, Nicola. And you should
too."

Nikki exhaled. Then she said with a grin, "Well, I'm
glad that's over with."

For several minutes they chatted companionably, but
as soon as Nikki saw her mother's eyelids drooping she
announced that she would have to be going. As she
gathered up her possessions, Nikki said in a voice which
successfully hid her resentment, "By the way, Mom,
I've moved out of Erika's and I'm staying with Michael.
I—uh—forgot to tell you on Friday."

"Yes, dear. He called me Wednesday and told me you'd be staying with him. He hasn't been to see me. You tell him I miss him," her mother complained jokingly.

Ice invaded Nikki's body. Wednesday? She had moved back to Michael's apartment on Thursday. He was so bloody arrogant, so sure of himself! Her tone was frigid as she told her mother, "He's away for the weekend. I'll tell him you asked for him—if I see him."

Mrs. Warren was too sleepy to notice her daughter's fury. Good-byes were said and Henry drove her home to the early dinner which Suzanna had expertly prepared. The meal looked delicious, but Nikki only sat and picked at it.

Chapter 6

Nikki walked out of Michael Cragun's bedroom with a satisfied smirk on her face. Her resentment of four hours before had dissipated. Now she felt only the expectant glee of the seasoned practical joker. If her imagination managed to think up enough such tricks, she might even begin to enjoy her Fifth Avenue incarceration.

She had placed a rubber snake she purchased the day before on one of the pillows, then carefully pulled the quilted bedspread back over it. It was impossible to tell that anything was underneath. Afterward she had taken time to look around a suite that appeared to have been designed so that its occupant would be self-sufficient in the event of nuclear attack.

Two sides of the room were windowed; a sliding glass door, identical to the one in her own room, led out to the terrace overlooking the park.

A room-sized walk-in closet with built-in drawer space held all of Michael Cragun's clothing. And beyond that was a dressing room complete with refrigerator, two-burner stove, and liquor cabinet. The bathroom held a large, deep Japanese tub that was easily big enough for two people. No doubt Michael took full advantage of that particular feature.

An hour later, after a soothing bath, Nikki turned on the television set and curled up contentedly on her bed. For the second time in four nights, she fell asleep in the middle of an old movie.

She was awakened from a trance-like sleep by a piercing shriek (female) and a gust of laughter (male). She sat up abruptly, pulling closed the furry pink robe that was her only garment. A sleepy hand brushed her long hair from in front of heavy-lidded eyes.

The door was opened, the overhead light clicked on. Michael Cragun stood before her, shirtless and shoeless, gingerly holding up the rubber snake in his left hand.

"Your latest pet seems to have slithered his way into my room," he told her indifferently.

Nikki, squinting from the bright light in her eyes, studied his face for signs of anger or annoyance. As much as she relished taking her revenge for the games Michael played with her, she was more than a little afraid of him too. Although his mouth was expressionless, she spotted warm amusement in his eyes.

"I suppose," she answered, mimicking his detached tone, "that he wanted to be among his own kind."

"Really! How perceptive of him to seek such superior company. And how disappointing for him to have to stay with you!" Michael shot back. "*You*, Miss Warren, are going to get up and apologize to Miss Dunne. She was so startled she may miss her high notes for a week."

Miss Dunne? High notes? Nikki couldn't deny that the tempestuous, red-haired soprano looked and sang like an angel. But she thought with disdain that the woman behaved rather like the reverse. No one could forget the stir she had caused in opera circles a few years before when she had insisted on playing a seduction scene in the nude. Although the lighting was dim, the male critics were effusive in their appreciation.

Now Nikki knew who Michael had spent the weekend with, but she had no intention of walking nearly undressed into his bedroom and apologizing to his paramour.

Michael saw the stony refusal written all over her face. "Get up and get it over with, Miss Warren, or I'll have to see that you do," he said wickedly, obviously enjoying the whole situation immensely. He would probably love it if she refused, thereby giving him an excuse to carry her bodily into his room.

That image was enough to make Nikki reconsider. She tossed her head, shaking her hair back from her face, and marched out of her room down the hall into Michael's. Elianna Dunne stood by the king-sized bed, clad only in a black nightgown, the open-work lace of which left little to the imagination. When she saw Nikki, whose bulky pink robe and long hair made her look an angelically beautiful seventeen, she raised an arrogantly inquiring eyebrow at Michael.

"She's—uh—a friend of the family. I'm watching over little Nikki while her mom's in the hospital," he explained, deadpan.

Little Nikki, indeed! She was tempted to pick up a pillow and heave it at him. Instead, Nikki lisped, "I'm *so* thrilled to meet you, Miss Dunne. And I'm terribly sorry that my little snake frightened you. It was just a private joke on Uncle Michael." She tossed a wide-eyed, innocent look at her supposed relation. "Are you sure you don't want to sleep with my cuddly pet? You'd get along so well together."

He handed her the snake, then drawled, "You sleep with the snake, little one. I'd rather sleep with Miss Dunne."

Outclassed, Nikki blushed and started to leave, but as she passed Michael she hissed in a low voice, "You're welcome to her, *Uncle* Michael."

He laughed and closed the door behind her.

Back in her own room, Nikki looked at the clock beside her bed. It was 10:36. She had slept for over an hour and was now wide awake. Her mind conjured up the shapely form of Elianna Dunne. Undeniably the soprano was as sensuous and sophisticated as she was talented, but Nikki found herself disliking the woman.

Nikki told herself that she should be delighted that he was focusing his attentions on the glamorous Miss Dunne. It meant he would leave her alone. And then a graphic picture of the two of them, entangled in each other's arms on his bed, floated before her, and she felt physically sick. She had never set herself up as a judge of anyone's morals. She recognized that Michael Cragun was hardly the type to be celibate. His personal life was none of her business.

Yet, lying on her stomach, her chin resting on crossed-over wrists, Nikki suddenly knew that she wished she could strangle Elianna Dunne, and that she would have felt the same way about any other woman who shared Michael's bed.

She mentally counted the number of times she had thought of him this weekend. She remembered her curiosity about where he was, and with whom. She recalled how shocked she was by her threat to abort his child. And worst of all, she admitted to herself that he had an overpowering physical effect on her, so that she felt like a babbling fourteen-year-old whenever he was too close.

You stupid little idiot, Nikki cursed herself, shaking her head. You've gone and fallen in love with a man so far out of your league that he thinks of you as the football!

She felt as cornered as a hunted deer. Her pride demanded that he never find out. She was certain he would be amused—and then proceed to do exactly as he wished with her. She knew of only one method of fighting back, and that was to be so thoroughly obnoxious to him that he would be happy to see the back of her.

She heard voices in the hall, footsteps on the stairs, the slam of the front door. Probably they had decided to go to Elianna Dunne's apartment.

Nikki was determined to avoid Michael as much as she could; it proved far simpler than she feared. Michael Cragun, she discovered, was a workaholic who spent fourteen or fifteen hours a day at his office. He usually left the apartment before 7:00 a.m. Nikki's job began at 8:30 so she simply waited until she heard him leave before venturing out of her room.

On Monday morning, she left the envelope he had sent to her office on his night table. Michael made no mention of her rejection of his gift, and when Nikki peeked into the room on Tuesday morning it was gone.

Suzanna prepared gourmet meals each evening and was insulted if Nikki failed to do them justice. As a result she gained four pounds by the end of the week. Each evening after dinner she disappeared into her room to read or watch television, lying in bed to listen for the sound of Michael's footsteps, for the casual slam of his bedroom door. Her heart pounded fiercely until she felt certain he had no intention of barging into her room. After all, he had done so on two occasions and apparently thought of himself as some local potentate who needn't bother to knock first.

Nikki found herself acting far more warmly toward the unmarried men with whom she worked, including her boss Ron Martens. In the beginning her smiles were pure feminine instinct, but she soon realized that her self-esteem would receive a much-needed boost if she were to be asked out by some of the men. She was confident that it would be easy to bring about.

Nikki was regarded by the men who knew her as rather unapproachable. She was not unfriendly, but her intelligence and beauty, combined with a wary aloofness, made most men feel that she was too good for them. Nikki was perceptive enough to realize this and was determined to change her image.

Unbeknownst to Nikki, she had set herself an impossible task. One of her co-workers had spotted her emerging from the sleek black chauffeured limousine; he had promptly asked a friend in the newsroom to find out who owned the car. The license plates read CAI INC, so it was hardly a surprise when the reporter's contact in the Motor Vehicle Bureau came up with the name Michael L. Cragun. As a result, Nikki's colleagues considered her Michael Cragun's very private property. As far as they were concerned, the lady was wearing a large sign reading "Do Not Touch."

To her dismay, Friday morning passed without an invitation from any of the men to whom she had been so resolutely friendly. She had begun to wonder what was wrong with her, and was reaching the point where she was ready to walk up to the nearest male to ask.

Earlier in the week she had made an appointment with her boss to go over several applications from candidates for sales jobs. Nikki told herself it was the perfect time to broach the subject of her own sad lack of popularity. She was uncharacteristically preoccupied as she sat in the comfortable chair on the other side of Ron Martens' desk. How was she going to phrase such a question? She couldn't focus on what he was saying, but stared out the window behind him at the dirty exterior of an old hotel.

After a few minutes Ron interrupted his own evaluation of one of the applicants to observe, "Nikki, you seem to be off in your own world. Is something the matter? Besides the fact that we have too many good candidates and not enough openings," he joked.

Nikki felt herself pale. Before she had time to change her mind she plunged in: "Ron," it was the first time she had called him by his Christian name, "is there something wrong with me? I mean, that men don't— like me?"

"What?" He was clearly astonished by her question. "You've got to be kidding. Half the guys in this office are crazy about you."

"That's not true," she mumbled, forcing herself to look into his eyes in order to judge his sincerity. "Nobody ever asks me out—not even to coffee or lunch."

Now her boss looked rattled. "You know how it is,"

he shrugged with obvious embarrassment. "Men don't like to poach on somebody else's preserve. Especially when the property owner is so—influential."

Nikki's stomach began to churn. "Would you please drop the metaphors and spell out exactly what you mean?"

He linked his hands behind his head and leaned back in his chair. His composure had quickly returned. "Sure. Why not? You get ferried back and forth from work in a limousine registered to Michael Cragun. Sam saw you and he had a friend downstairs check the plates. Frankly, everyone around here assumes you're his mistress." The tone implied that he found this distasteful.

"Well, I'm not!" Nikki denied heatedly, absolutely furious with Michael for this latest interference in her life. "I'm staying in his home—and it's—it's none of anyone's business why—but I don't sleep with him. Not ever! I haven't even seen him since last Sunday. I live my own life, and I'm no one else's *property!*"

"My apologies, Miss Warren," Ron told her with a smile. "So how about dinner tonight?"

Nikki had cooled down and was able to respond lightly, "Yes, Ron, I'd like that."

Nikki almost forgot to notify Henry not to pick her up. Suzanna took her message, telling Nikki to enjoy her dinner and not to worry—the steak for that night could be frozen. She was only concerned that Nikki might need a ride home later that evening, and explained that "the boss" would be at a charity dinner and Henry would have to wait for his call. Nikki assured Suzanna that she could take a taxi, and breathed a sigh of relief. *He* would never even find out that she had gone to dinner with Ron.

They went to a Japanese restaurant that was well

known for its authentic, expertly prepared food. After a glass of white wine, Nikki screwed up her courage and sampled some raw fish, which was served with a sauce so spicy that even the few drops she put on her fish burned her mouth. The sea bass itself was mild, the texture crisp. After a soup with a fish-stock base and a vinegary Japanese salad, they shared some shabu-shabu: raw meats and vegetables cooked in a broth at the table.

Nikki found Ron to be a compatible dinner date. He was physically attractive and easy to talk to. The two of them lingered over their dinner, sipping wine and sharing conversation.

Ron insisted on accompanying Nikki back to Michael Cragun's penthouse apartment. He had invited her to his westside brownstone for coffee, but she had politely declined. She was in no mood for fending off male advances. She had not intended to ask him in either, but when he kiddingly requested "the 25¢ tour," she could hardly refuse.

After a cursory inspection of the downstairs, Ron sat down on the living room couch and pulled an unresisting Nikki down to join him. No one seemed to be home. She began to tell him about the archaeological expedition that had discovered the primitive statue perching menacingly on a pedestal near the couch where they sat. Ron put his arm gently around her shoulder, saying huskily, "I really don't want to talk about some millionaire's art collection. In fact, I don't want to talk at all." He began to kiss her, carefully and slowly.

The sensations he aroused in her, while not intensely exciting, were agreeably pleasant. He was certainly experienced with women and considerate of her feelings. Nikki liked him a good deal. She found herself wanting to experiment, to determine if she could respond to Ron as she had several times responded to Michael.

She twisted around to face him, her hand resting lightly on his cheek, and began to kiss him back. His lovemaking was tightly controlled, as if he were afraid of frightening her off. But as the minutes passed, his kisses became increasingly passionate and rough. His hands bit into her skin, holding her tightly; she could feel his heart beating rapidly against the thin material of her blouse. She could tell from his uneven breathing that he was far more aroused than she was, and when his hand slipped under her dress and began to inch up her thigh, she stiffened. She slipped one palm between them and tried to push him away.

Ron only tightened his hold on her, and with his lips still against her mouth mumbled, "Come home with me Nikki. Please."

"Ron, no—" she protested, but managed only these two words before he clamped his mouth over hers again. She was not at all frightened—after all, what could happen in the middle of Michael's living room?—but she knew that she wanted him to stop. What had been welcome before had become offensive.

Nikki was trying unsuccessfully to extricate herself when a voice drawled from behind them, "I think it's about time the wrestling match ended. I call this round a draw."

Ron Martens reluctantly drew away. Skin damp and tie askew, his appearance presented a sharp contrast to that of Michael Cragun, who was leaning negligently against the tall arched entryway between the hall and living room. He was dressed in formal attire, but his tie had been unknotted and hung carelessly around his neck. One elegant shoulder rested against the door frame, his left leg lazily crossed over the right. He gave the impression of a man who had been standing and watching for quite some time.

How much had he seen? Nikki hadn't heard the front door open; he must have used the service entrance off the kitchen. She stared at the intricate design of the oriental carpet beneath her feet.

Michael strolled casually over to them and half sat, half leaned against the arm of an upholstered chair diagonally adjacent to where Ron was sitting. "May I ask whom I have the—pleasure—of meeting?" he said to the man next to Nikki.

"Ron Martens, Mr. Cragun. How do you do?" They coolly shook hands, ignoring Nikki completely.

"Mr. Martens. You're Nikki's boss, is that correct?" Michael asked conversationally.

"Yes." Ron's tone was wary.

"In that case," Michael went on blandly, "I strongly suggest you restrict your dealings with her to the office."

The order was blatant provocation and Ron Martens made the tactical error of rising to the bait. He said angrily, "I think Nikki is capable of running her own life, Cragun. And I damn well *know* I am!"

"I agree. *You* are. But she's not. I'm responsible for her and she'll do what I tell her," Michael said in his haughtiest lord of the manor fashion.

Nikki's eyes jerked up from the floor to glower resentfully at Michael. "You are *not* my keeper, Mr. Cragun. I'll date whom *I* choose. I don't let you tell me what to do."

"Don't you?" he needled. He stood up, stretched. "Okay, children," he said with a yawn, "time to break this up."

Now Ron Martens exploded. "Cragun, don't talk to

me that way," he shouted. "I'm older than you are, for God's sake. And Nikki will make up her own mind—"

"Please, Ron, you'd better go," Nikki quickly interrupted. She was more sensitive to Michael's mood than her livid boss was. Michael might look outwardly relaxed, but there was a slight downward curve to his mouth and a narrowing of his eyes that told her he was on the verge of losing his temper. She could picture him grabbing Ron Martens by the collar of his shirt and literally throwing him out the door.

Michael cast an imperious eye on Nikki. "An excellent suggestion, Miss Warren," he said sarcastically, and then looked coldly at Ron. "But in case my feelings aren't clear, Martens, let me spell things out. I want you out of my home. I *don't* want to see you here again."

He tossed Ron Martens his suit jacket and overcoat, which had been thrown over the back of the couch. It was a gesture of dismissal.

Ron said an unsmiling good night to Nikki and walked stiffly toward the door. She attempted to follow—to offer a quiet apology—but Michael's firm hold on her upper arm prevented her from moving. He told her in a low voice, "You're not going anywhere."

Nikki waited until she heard the slam of the front door. She had been able to hold her temper in check only because she feared that making accusations in Ron's presence would have inflamed him into trying to fight Michael. She had no illusions as to who would have wound up out cold on the expensive oriental carpet.

But now they were alone and her rage boiled over. She yanked her arm away and backed off from Michael before she could succumb to the futile urge to slap him. Then she mounted a blistering verbal attack. "You filthy pig! Keep your dirty hands off me! Where do you get off interfering in my personal life?"

He lectured her as one would a fractious child. "Just calm yourself down, Nikki. I'm not going to argue with you tonight. You're too worked up. We'll discuss it—"

"Now!" she interrupted, her voice at a screaming pitch. "Just what were you doing, watching us that way? I think you're sick, you know that?"

"Okay," he replied in a tightly controlled voice. "Have it your way. We'll talk now. I witnessed your little love scene from start to finish. I thought you might need some help. True, you seemed enthusiastic enough at first, but when Martens' hand began to wander up your leg, you certainly stiffened up fast. The guy was trying to rape you in my living room, and you're mad at *me?*"

"He was hardly *raping* me," she snorted disdainfully. "I can take care of myself. I don't need you to hover around. Besides, it's none of your business what I do."

He forced her hand down before it could connect with his cheek. "You never learn, do you?" he said dangerously, his anger all too apparent now. As if to underscore his superior strength he twisted her arm into a painfully unnatural position. Nikki was too stunned by his brutality to move or answer back.

Michael said coldly, "You don't go out with anyone, you understand that? Any child you have is going to be mine, no one else's." His eyes roved over her body in a deliberately insulting manner.

When Michael released her wrist to grip her shoulders and propel her toward him, she panicked. As a sophomore in college Nikki had taken a course in self-defense. Now she applied what she had learned. In the wake of her near-paralysis, the abrupt attack caught Michael totally off guard. Her knee shot out and struck him hard in the groin, causing him to double over, obviously in agony.

"Michael?" Nikki's fear and fury of a few minutes before subsided. She was terrified she had inflicted some permanent injury. "Michael, are you all right?" she whispered.

He slowly straightened up, a taut smile forced into place on his white face. "I'm beginning to think I underestimated you. Maybe you're right—you can take care of yourself. You'd probably have made mincemeat of Martens."

The next morning Nikki hesitantly walked into the kitchen for breakfast to find Suzanna Merola making mushroom omelets while her husband and her employer sat at the breakfast table, the morning *Sun* in disarray before them.

Reassured by the normalcy of this scene, she smiled good morning to Suzanna and approached the table, overhearing the end of their conversation.

". . . and check on the liquor, would you Henry?" Michael was saying. "I don't know what Suzanna's planning. I'll leave the choice of wine to your impeccable palate." He turned to Nikki, who stood uncertainly a few feet from the table, looking coltish in blue jeans and a pink short-sleeved sweater. "Good morning. You *are* the lady with the bionic knee?" he questioned wryly.

Her voice faltered. "I'm really sorry, Michael. I—"

"Forget it. Maybe I deserved it," he shrugged. He got up to formally pull out a dinette chair for her. "We're going to the hospital after breakfast to see your mother."

Nikki immediately took umbrage. "Is that another one of your orders, Mr. Cragun?"

With a grin, Michael put out his hand to shield himself from her knee. "Forgive me, ma'am. I'm going to have to learn to be more polite with you, hmm?"

"It would be a nice change," Nikki agreed sarcastically.

"Right." He cleared his throat and affected a snobbishly formal accent. "Would Miss Warren honor us with the pleasure of her company tonight at a small, very exclusive dinner party?"

"Here?"

He reverted to New York-ese. "Of course, here. I'm not a total pariah, Nikki. I do entertain—more than one person—at times."

His gentle dig brought to mind the unpleasant memory of her encounter with Elianna Dunne. She sincerely hoped that the sensuous soprano was performing in a very long opera tonight. But Nikki was not about to make the mistake of pressing Michael for a recitation of the guest list. He had adopted a teasingly winning manner with her and her world lit up.

"I'll look forward to meeting your friends, Michael," she dimpled, "almost as much as I look forward to Suzanna's dinner. And if there's anything I can do—"

"There is," he said seductively. "You can sit down and eat your breakfast. You can visit your mother with me. And tonight you can make yourself look even more beautiful than you do now." He smiled. "Please."

Look at me like that, she thought, and I'd crawl on my knees to China for you.

Michael drove the Cadillac to the hospital, over the objections of an unhappy Henry, who complained that Suzanna would find some disagreeable job for him to do if he stayed home. As they rode up to the Bronx, Nikki informed Michael that Mrs. Warren had been grousing since Sunday because he never came to see her. "She says she misses you. God knows why, but you made a big hit with her."

"I'm glad one of the Warren women appreciates me," he teased. "But I do keep tabs on her. Steve Rowland is a friend of my father's. He's the one who did the heart surgery a few years ago. You know who I'm talking about?"

Nikki nodded. "I saw him holding mom's hand. He seems very nice."

"He is." Michael glanced at Nikki, switched on the radio, then abruptly turned it off again. "Look, Nikki. About last night—"

"Do we have to talk about that?" she asked anxiously. The last thing she wanted was another verbal battle. "I'd rather just forget it."

"Yes, we have to talk about it," he insisted. His voice became persuasive. "Relax, honey. I'm not going to bite your head off again. I *do* deny that I'm a filthy pig. Or some kind of voyeuristic sickie. But about the rest of it, you were right. I came on too strong. You want to go out with Martens, go ahead. It's not my place to stop you, although I think you can do a helluva lot better. I want you to feel that my home is your home. I had no right to take over your life and now the least I can do is look after you. It's become a habit." He smiled charmingly at her.

Nikki was momentarily astonished by this complete about-face. Then her defenses slammed into place and she said mistrustfully, "Don't tell me. You've decided you don't need a female—you're going to get yourself cloned."

He began to laugh. "I know you think I'm arrogant, but even I wouldn't go that far. Fatherhood has lost its appeal. Just think of me as a big brother, Nikki."

Nikki could not have been more dismayed if he had announced that he was throwing her out in the street.

Was that how he saw her? As a little sister? She decided that it must be so. He had never touched her, except on those occasions when she had strongly provoked him. Evidently he preferred older, more sophisticated women. She had been living with him for over a week. If he had wanted to seduce her, all he had had to do was walk across the hall.

She should feel like Cinderella on her way to the ball. Her pressing financial problems were solved; she could pay Michael back at her convenience. She had the use of a beautiful apartment and a chauffeured car, and her mother had improved so rapidly in the month since the second stroke that she would soon be able to leave the hospital. Best of all, she was free to do as she pleased, without the threat of interference from the dictatorial Michael Cragun. But unhappily for Nikki, she was so in love with him that this brotherly solicitude was a form of torture.

Michael was charming and attentive to Mrs. Warren during their visit. She had smiled delightedly upon spotting him in the hall outside her open door and insisted that he sit by her bed. A disgruntled and ignored Nikki stood by the window, half-listening to their conversation. Michael told her mother he would transfer her to the rehabilitation center as soon as he could badger the doctors into agreeing. The March Institute was in Manhattan, some thirty-five or forty blocks from his apartment. He promised to visit more often once she was in residence there.

If Pam Warren noticed her daughter's listlessness, she said nothing. As Michael summoned Nikki to leave, Stephen Rowland came in. As Michael took her arm and ushered her out, he looked back over his shoulder and said to the doctor, "If you think the mother's a challenge, you should try the daughter."

Michael turned the full brunt of his considerable charm on Nikki during the ride back. Although she wondered why he was bothering to be so pleasant, she permitted herself to be teased into a conversation. They were talking easily as they entered the apartment building.

Suzanna greeted them with a thumbs up signal as they passed through the door.

"I have a surprise for you," Michael announced, looking very pleased with himself. "Close your eyes."

Although puzzled, Nikki obeyed. Her hand was held in Michael's as he led her through the hall into the living room. "Okay, you can look now."

In a previously empty corner of the room stood a highly polished piano: a baby grand Steinway similar to the one Nikki had admired in Florida. "When— where—?" she laughed.

"Promise me you won't try to bite my head off if I tell you!" Michael demanded with a rakish half-smile.

"Yes—I mean no—"

"I bought it the day after I met you. And if that means I made up my mind to have you live with me, then I did—so don't frown at me. The piano belonged to a friend of—uh—Elianna's. A singer and composer. He decided to get himself a concert grand. His piano arrives Monday; I had to bribe the piano movers to get this here this morning." He was absolutely delighted with himself.

Nikki's expression registered her amazement. "But why?"

"Never let it be said that I don't want to keep you happy, little sister," he said evasively. "Now why don't you play something soothing while I go into the study to work, hmm? I could use some inspiration." He flipped up the piano bench, revealing Nikki's music.

She selected a book of Chopin pieces and began to play.

Michael's guests were invited for dinner at 7:30, but this was New York and people were fashionably late. At 7:35 there were two staccato knocks on Nikki's door. She called "Come in" as Michael, wearing a beautifully tailored suit, his jacket slung over one shoulder, walked halfway into the room. He sat down on her bed, and watched as she brushed her already smooth hair.

"See how well-trained I've become? I even knocked," he boasted, as if proud of some major accomplishment.

"But you didn't bother waiting for me to invite you in before you made yourself at home," Nikki retorted.

"Ah, but it *is* my home," Michael needled gently.

Unwilling to continue the war of words, she silently began to pin her hair up. "No, don't do that," Michael ordered with an unapologetic return to his accustomed lordly manner. "Wear it the way you did in Florida, with that flower thing. It goes with the dress." Nikki was wearing the sunshine yellow and orange halter-necked dress that she had acquired in Florida. She tried to decide whether to accede to this dictum. On the one hand, she was flattered by his interest in her appearance and wanted to please him. But she bristled every time he gave her an order. Their eyes met in the mirror and Nikki quickly looked away to continue her absent-minded brushing.

Michael walked over and impatiently removed the brush from her hand, smoothing her hair with his palm as if she were a little girl. "Come on. Stop primping. You already look perfect and you have to play hostess."

Nikki melted at his touch, her agitation with his high-handedness quickly forgotten. She pinned the silk flower into her hair, glowing from his compliment and pleased by his assumption that she would act as the hostess tonight. They descended the stairs, Nikki on Michael's arm, as Suzanna watched.

To Nikki's relief, she had already met two of Michael's guests—Peter Delavan and Charles Morris—although in admittedly dubious circumstances. Peter and his vivacious wife Samantha were the first to arrive. Samantha was a sophisticated redhead with short curly hair and an advanced case of pregnancy. The Delavans were followed five minutes later by Melanie Cragun Newman and her husband Brandt, both of whom looked healthy and tan after ten days in the Caribbean and Florida. Unlike her often intimidating younger brother, Melanie Newman was sociable and warm. Upon being introduced to Nikki, she said enthusiastically, "Mom told me all about you when Brandt and I were in Florida!"

Nikki busied herself passing out hors d'ocuvres while Henry mixed drinks. But soon she acceded to Sammi Delavan's insistent request that Nikki join her on the couch. Sammi made no attempt to disguise her curiosity, telling Nikki blithely that she was dying to find out how Nikki had "hooked" Michael when half a dozen of her friends had tried and failed. Nikki had just mumbled that there was nothing official between them when Charlie Morris, Michael's attorney, and his wife Sheila came in. As soon as Sheila had excused herself to comb her wind-blown hair, Michael all but dragged the lawyer over to the couch and pointed to Nikki. "See, Charlie? Alive, well, and unharmed. And I promise you as pure and virtuous as ever. Now will you get off my back?"

"It was only four phone calls, Mike. Just keeping you honest," the older man said with a laugh. He walked over to the bar for a scotch and water.

Nikki's face was still pink with embarrassment as Sammi Delavan subjected her to a thorough visual dissection, ultimately drawling, "I must tell Craig to give Michael a complete physical examination. I've never known him to be honorable before. So he must be sick," she paused to sip her drink, "or in love. And speaking of Craig—" she nodded toward the door at a stocky, slightly disheveled man who had just come in with a freckled blonde.

Nikki was soon being introduced to Michael's college roommate Craig Landesman, and his date Merry Quinn. They were both doctors specializing in pediatrics, and Michael endured with good grace the usual ribbing about it being time to avail himself of their services. Charlie Morris caught Nikki's eye, noting with a frown her quick blush.

Suzanna had produced a five course French dinner so superb that it would have compared favorably with the fare at La Caravelle or Lutèce. They ate with slow appreciation, drinking a different wine with the seafood crêpes, cream of cucumber soup, and stuffed shoulder of veal. Nikki couldn't manage more than a bite of her spinach salad, and felt like a glutton by the time she had polished off a large serving of the Soufflé Grand Marnier.

Nikki had been expecting serious dinner time conversation, perhaps economic shoptalk, but in fact she laughed so much during the meal that her stomach ached more from that than from the amount of food she had consumed.

When Michael had taken her arm to seat her at the opposite end of the table from himself, she had hoped that her self-consciousness was not obvious to the others.

After dinner Michael ordered her to the piano

whispering that she could always start with a love song again. Nikki ignored the teasing. Since that afternoon she had been eager to play again.

Her solo soon turned into a sing-along with Michael suspiciously silent. Although he tried to pass himself off as "the audience," his sister Melanie goaded him so relentlessly that he finally joined in on a chorus of an old folk song. His less than melodic efforts were greeted by hoots of laughter from his friends. When his face actually reddened in response to everybody's teasing, Nikki was totally delighted. She hadn't believed the arrogant Michael Cragun to be capable of embarrassment.

The ringing of the doorbell at 11:30 went unnoticed by those gathered around the piano. But Elianna Dunne, in a flame-colored dress, made sure that her entrance would play to a full house. Her loud, "Michael, darling," was no stage whisper, and as she approached her target she dramatically threw her arms out to entwine them around his neck and kissed him passionately on the mouth. "I'm so sorry I couldn't make it earlier, mon brave," she said huskily. "The curtain calls went on forever tonight!"

Nikki's playing became mechanical. So she was just a substitute for the prima donna after all. Her whole evening, which up until now had been enchanted, was spoiled. Michael and Elianna immediately disappeared; as soon as Nikki finished the song she was playing, she excused herself as well. She was too miserable to notice the sympathetic looks from the others.

She climbed the stairs slowly, overwhelmed by total dejection. Of course she had known that Michael's solicitousness and warm glances were meant platonically, but it had been so easy—and pleasant—to imagine otherwise. An angry female voice emanated from behind Michael's closed bedroom door. Ashamed of eavesdropping but unable to help herself, Nikki stood in the hall and listened.

". . . made her sound like a pimpled adolescent, and heaven knows she looked like a child the other night. How old is she, Michael?" Elianna Dunne sounded furious.

"Twenty-three." His tone indicated that Nikki was of no importance.

"What is she doing here? She's no 'friend of the family,'" the diva spat out.

"Ellie, you're making a scene," Michael said evenly, "and I won't tolerate that. My friends—"

"To hell with them! That—that—child! Oh my God, Michael, this doesn't go back to that stupid argument we had about children?" Her voice became sultry, pouting. "Darling, you didn't take it seriously when I dared you to find somebody else to have your children?"

"I believe the word you used was brats, Ellie," he told her icily. "Now back off. You're getting your makeup all over me." Nikki thought to herself that if Michael ever used that tone on her she would want to crawl into the nearest hole.

But the lovely Miss Dunne was undeterred. She purred, "Don't be cross with me, cheri. You can't mean to use that skinny creature to—"

"No," he cut in. "She's just a child herself. Probably needs to be taught where the little angels come from in the first place." He paused, and when he spoke again his voice was harsh. "There's a certain type of female, Elianna, that bores me senseless. Let me tell you something about Nikki Warren. She—"

Nikki did not wait to hear the rest. She fled down the hall into her room, slammed the door, and flung herself onto the bed. Emotions that had been see-sawing wildly for the past twenty-four hours finally cracked under the strain. Now she knew Michael's true opinion of her. She was not only an ignorant child, but even worse, she bored him to death. She had no sparkle, no excitement. She would never be provocative and seductive like Elianna Dunne.

Tears filled her eyes and soon she was sobbing uncontrollably. She cried until she had so exhausted herself that she had no strength left to continue. Zombie-like, she shrugged off her dress, leaving it in a heap on the floor, and fell back into bed.

Chapter 7

Unwilling to risk seeing Michael, Nikki hid herself in her room the next morning. It was nearly noon when Suzanna came in carrying some coffee and toast on a tray. The housekeeper assumed that the boss's inexperienced young guest had a bad hangover. One look at Nikki's puffy eyes and drawn face told her otherwise.

Suzanna said nothing to Nikki, however, but simply smiled sympathetically and urged her to try to eat something. Alone, Nikki nibbled the toast and wondered if it had been only a week since she had realized her feelings for Michael. She admitted that his wealth, suavity and charm had resulted in almost instant infatuation. But it quickly deepened into love in Florida, when she was exposed to his sharp, penetrating intelligence. Even the darker side of his nature—his arrogant manipulation of her actions, his use of raw masculine sex appeal to suit his own ends—held a frightening fascination for her.

She cursed her failure to stick to her original intention to be sarcastic and unpleasant. Short of affecting absolute indifference—an acting feat far beyond her meager talents—surely icy disdain would persuade Michael that the game had palled. Nikki would not admit to herself that such behavior was really an attempt to provoke him into betraying some response to her.

She successfully avoided him until late that afternoon, when he walked into the living room as she sat playing the piano. Dressed in a Yale tee shirt and gym shorts, sweaty from his handball game, he stood and listened while she finished the Gershwin piece she was practicing. She accepted his praise mutely, her temper beginning a slow burn when he said in an earnest tone, "We all missed you last night after you went to bed. I'm sorry you weren't feeling well. Next time I won't let you drink so much."

"I thought we agreed that you weren't my keeper, Michael," she said with cloying sweetness. "Besides, I was only the second string. When the first team arrived, I decided it was time for a strategic retreat."

Michael sat down next to her on the piano bench. "You want to tell me what the hell you're talking about?"

"You figure it out!" she snapped, and stormed out of the living room. Ten minutes later she heard the front door slam. Michael did not show up for dinner.

From the moment Nikki got into bed Sunday night she had been dreading Monday morning. She had to meet Ron Martens sooner or later, and she knew that it would be awkward for both of them. Her boss, however, had not reached the age of thirty-seven without knowing how to gracefully extricate himself from a love affair gone wrong. Toward Nikki he was business-like but nothing more. As soon as she realized that there would be no embarrassing confrontation, she relaxed. Nikki resumed her previous aloofness to the other men in the office. Now that she was officially free to date whom she pleased, she found that she had no desire to do so.

Erika was taking a week's vacation in order to attend her brother's wedding. Nikki threw her energies into the additional work created by Erika's absence. It kept her mind from wandering to more painful subjects. Some employee records had gotten hopelessly mixed up; it was a challenge to sort them out.

Michael was home by seven o'clock Monday and Tuesday evenings. Nikki would have preferred that he work late. She purposely turned their meals together into a cold war. On Monday he had acted as though they had never quarreled, smiling hello to Nikki and telling her about his problems in finding a capable temporary replacement for his very efficient secretary. Nikki had stared into space as if dreadfully bored. "Okay, I get the message. You're sick of hearing me talk shop. I don't blame you," Michael admitted. "Tell me how things are going at the *Sun*."

Nikki's cool reply—"I'm not really interested in telling you anything, Michael"—had not deterred him. He ignored her petulant mood.

However, when she seemed just as indifferent to his suggestion that they visit her mother and was noncommittal toward an invitation to attend the theater later that week, he said with some irritation, "I'm not a mind reader, Nikki. I have no idea what sin I'm supposed to have committed. So enlighten me."

She replied, "Frankly, Michael, you've just ceased to interest me, that's all." She added a yawn, thinking it a nice touch. Michael shrugged and concentrated on his roast duckling.

The next evening as well, Nikki lost no opportunity to be coldly and deliberately nasty. Several times Michael seemed to be on the verge of losing his temper but he never gave her the satisfaction of doing so. The closest he came to registering annoyance was after dinner, when he told her, "I won't be home for dinner tomorrow night. I have a meeting. I'm sure you'll be disappointed that you won't have Michael Cragun to kick around anymore." He smiled as he spoke, but his eyes were blue chips of ice. Nikki stared back just as coldly, but inside she was quivering.

As Nikki left the *Sun* building Thursday afternoon, a feeling of unease stole over her. Henry was not parked in front of his favorite fireplug. Nikki was standing outside the building trying to decide whether to wait a while or take a bus when the black Cadillac pulled up beside her. In the back seat were Michael Cragun and two dark-suited men.

She half-expected Michael to totally ignore her, but instead he opened his door, bowed mockingly, and needled, "You have your choice. You can either sit in the front with Henry or come back here and sit on my lap."

"No contest, Mr. Cragun," she sniffed as haughtily as she could manage. "I'll pick Henry's company over yours any day of the week." She opened the front door and slid gracefully into the seat. Nikki had noticed several times before that the chauffeur had the knack of seeming to become completely invisible when the situation warranted, and he had just pulled this remarkable disappearing act. He did not say hello to Nikki. An android might have been driving the car.

Michael slid open the smoky glass panel between the seats and turned to the man next to him. "My irresistible charm seems to have met an immovable object by the name of Nikki Warren," he noted drily. "Could you bring yourself to turn around, ma'am, and meet the hired help?"

This gentle teasing tempted her to melt and smile at him, but she resisted that impulse. Instead she stared straight ahead at the rush hour traffic and replied jeeringly, "No thanks. If they work for you, I'd have to question their integrity."

She felt a hand on her shoulder and struggled not to flinch. Michael's voice next to her right ear was seductively coaxing as he told her in a low tone, "I know you're angry with me, and I know you don't feel like telling me why. But please don't involve my friends in the hostilities. Come on, honey; turn around and say hello."

His endearment infuriated her. She wasn't his honey; on the contrary she was inexperienced, ignorant, and boring! She lifted her chin, straightened her shoulders and directed evil thoughts toward Michael Cragun.

"Nikki?" He put his right arm along the back of the seat, resting his chin on his wrist. He allowed his left hand to drop down to her shoulder and wander over to her jaw, gently forcing her head sideways to nuzzle her cheek.

Angry with herself for being aroused, Nikki pulled away and twisted around, intercepting the glances of the other two men, who regarded her as one might a petulant child. Nikki forced herself to smile sheepishly. "I'm sorry for being so moody. Bad day at the office," she lied. "It's nice to meet you both."

"Jack Wright on the left and Shaun Bernstein on the right," Michael told her. He was once again sprawled negligently back in the rear seat.

Nikki was unable to suppress a grin upon hearing the second name. Then she reddened. It was one thing to be deliberately nasty to Michael; another to laugh at a total stranger.

But he seemed to read her mind. "Irish mother and a Jewish father. I'm married to an Italian girl and our first baby's due in May. The family has a contest going to see who can think up the best name."

"How about Arabella Bridget Bernstein," suggested Nikki impishly.

"Carlo Kevin Bernstein, if it's a boy," chimed in Jack Wright.

During the remainder of the trip, the three of them took turns calling out improbable names. Their laughter was uninhibited and frequent. Michael Cragun did not join in; he sat and stared out the window, his face unreadable.

"You'll join us for dinner." As on so many other occasions, it was an order, not a request. A streak of rebelliousness incited Nikki to refuse, until she realized she could exasperate Michael far more by accepting.

"Of course," she said sweetly, "I wouldn't pass up the chance of getting to know your charming colleagues." She flashed the men a lovely, shy smile, then murmured in a husky voice that she would be back as soon as she changed into something more comfortable.

She sashayed into the living room wearing a green and white jumpsuit, long-sleeved, high-necked and clinging. Her hair hung in careful disarray down her back. She was woman enough to revel in the obvious admiration of Michael's two employees, who stood with their boss in the living room, sipping martinis.

Michael asked politely, "What can I fix you to drink, Nikki?" His voice was bland and his appraisal of her figure unimpressed. Nikki resented his indifference and coldly refused his offer.

She opened her green eyes wide and gazed up at Jack Wright. Nikki had learned while waiting on customers at her mother's store that people tended to buy more when subtly flattered. She employed her expert technique on the unsuspecting bachelor at her side. He was convinced that the beautiful young brunette found him absolutely fascinating. Unfortunately she was clearly the boss's property, and judging from the look he'd surprised on Michael Cragun's face a moment ago, she had better take care. Then Jack Wright mentally shrugged. He couldn't help it if the lady found him appealing.

When Michael excused himself to make a phone call, Nikki mixed herself a gin and tonic. Then she focused her charms on the very vulnerable Shaun. By the time Michael returned to the room, she knew a good deal about both Jack and Shaun: their jobs, their families, their education.

This was a working dinner and conversation naturally concerned CAI. Much of the jargon-filled talk was incomprehensible to Nikki; she had to content herself with admiring glances and smiles from the two susceptible males on either side of her. Michael, who sat across from her at the head of the table, she studiously ignored.

After dinner the men rose to closet themselves in the study. Nikki murmured the usual good night pleasantries. "I hope your wife feels better for the rest of her pregnancy, Shaun." She studied his face winsomely. "Personally, although I think it's fine for a man to be successful in business, what really impresses me is a devoted husband and father."

Jack Wright asked quickly, "Are you taking applications?"

Nikki made her reply seductive. "Ask me out to dinner and you'll find out."

Jack's glance darted to his boss. Michael was leaning against the dining room table, his arms folded across his chest. He stared unsmiling ahead, a hard expression in his eyes.

An awkward silence filled the room. Nikki broke it by cocking her head toward Michael and drawling, "I see the corporate giant is impatient to get to work. Good night, Shaun, Jack." She sauntered gracefully from the room, gently flipping her hair off her shoulders with a sensuous motion of her hands and a toss of her head.

All day Friday the sky had been oppressively bleak with heavy gray clouds. As Nikki emerged from the elevator she could hear that the not so gentle spring rains had finally begun to fall. She stood in the lobby and looked out the glass doors at the torrents of water streaming down. Passers-by either scurried through the rain, newspapers held above their heads, or took shelter in doorways. Nikki fished in her purse for an old rain bonnet she carried around for such emergencies. It was plastic, imprinted with the facsimile of the New York *Sun* headline for July 21, 1969. MEN LAND ON MOON, it proclaimed. ARMSTRONG AND ALDRIN EXPLORE WHILE COLLINS WAITS IN SHIP.

Nikki wistfully remembered how she and her parents

had stayed glued to the television set that evening,
waiting for the first live pictures of men walking on the
moon. Theirs had been such a happy, close family.

A rude honk brought her out of her reverie. Nikki
looked up, startled, to see that Michael Cragun was
again sitting in the sleek car as it idled, waiting for her.
As she dashed through the downpour, he got out,
protective umbrella in hand, and held the door open for
her. He gripped her arm tightly, pushing her into the
back seat next to him.

His brief touch was both exciting and painful. Nikki's
emotions had been heightened by her reminiscences in
the lobby, and she avoided looking at him. She knew he
would read uncertainty and vulnerability in her eyes.
She removed the wet rain bonnet and tossed it on the
floor. Henry, on the other side of the closed partition,
concentrated on his driving and was as invisible as the
day before.

"Quite a little performance you put on last night.
What were you trying to prove?" The tone indicated
mild curiosity, no more.

"Nothing." Nikki looked out the window. In spite of
the heated warmth of the car, her whole body felt
chilled. She knew she had goaded Michael relentlessly
yesterday, and was afraid of the punishment he would
exact. She tried unsuccessfully to suppress a sudden
shiver.

"I know the weather is fascinating, but would it
offend you to look at me when I talk to you?" Michael
sounded mildly peeved.

Nikki turned slowly toward him, her body stiffly
huddled into her coat, her arms tightly crossed in front
of her. "I have nothing to discuss with you." Although
she held herself rigidly in check, she was trembling
uncontrollably.

Michael took in her chattering teeth and clenched

arms. "Are you cold?" he asked with concern. He leaned forward, opening the partition momentarily. "Henry, will you turn up the heat, please?"

Nikki muttered her thanks. "Hey," he exclaimed softly. "Let me warm you up." He moved closer and started to put his arm around her shoulder.

Her objectionable manner of the last week became impossible to sustain. His gentleness sabotaged her bad intentions. She jerked away, choking out, "Don't touch me! I've told you I can't stand it!"

Michael uttered an expletive. "You're totally neurotic, you know that? I took a hell of a lot from you last night. You were unforgivably rude to me in front of my employees, and then you came on to them like a Forty-second Street hooker. It was all I could do to get their minds off your body and back onto business. Jack Wright is probably still panting. Do you get a kick out of being a tease? Is that it?"

"I'm not," Nikki mumbled.

"Like hell! For the last week you've provoked me every chance you've gotten. But every time I come near you, you flinch away like I'm Count Dracula. Make up your mind, Nikki. Yes or no?"

He looked coldly over at her until she cringed. She could feel him studying her, and was stunned by her physical reaction to his angry scrutiny. It excited her. She desperately wanted him to make love to her. She tried to dismiss the ache, the shortness of breath, the heat, as so much neurotic masochism. The man was so arrogant that he considered it an honor for a woman to be permitted in his bed.

Nikki stared down at her folded hands as he continued to watch her. And then the badly-cracked dam inside her crumbled and gave way. She swung about to face him, her eyes unable to meet his, and mumbled, "Yes."

For a long moment Michael said nothing, made no move to touch her. The implicit rejection mortified Nikki. She was about to retreat to her corner of the car when he demanded, "Look at me."

She obeyed, her face pale, her eyes defensive. He continued coolly, "Take off your raincoat. I don't feel like getting soaked."

Again she complied, tossing the coat onto the floor to join the still-dripping plastic bonnet. She looked up at him warily, awaiting the next set of instructions.

"Very good," he said softly. "Now your hair. Pull out the pins. Slowly. Let's see if you can be as seductive as last night."

Nikki had never been one to flee in the face of a challenge. A little smile pulled at the corners of her mouth. "More of your games, Michael?" she purred in her best temptress tone, almost enjoying herself. She languidly carried out his command, flicking the discarded hairpins into his lap. Then she combed her newly freed hair with her fingers, folded her hands primly, and sat motionless, the picture of docile submissiveness.

"A+, lady," he grinned. "Lesson number two—put your arms around my neck—like this," and he took her wrists and twined them together behind his head, "and tease me a little, Nikki. Make me want you so much I can't stand it."

But Nikki had a mind of her own. She did not produce the feather-light kisses she knew he expected, but simply nestled close to him, her eyes downcast, her hands unlinking to stroke his hair and face. In answer he slipped a hand under her blouse to caress her breast.

This silent battle was terminated by mutual consent. One moment they were staring at each other, the next kissing with the hunger and naturalness of reunited lovers. But soon the initiative was all on Michael's side. His rough exploration of her mouth had nothing in common with his deliberately persuasive lovemaking of a few weeks before. His hands and lips were wild, uncontrolled, and Nikki could only cling and respond. Minutes later he pulled his mouth from hers to bury his face in the hair which flowed down her neck and shoulders. He began a none too gentle nibbling of her soft neck. Nikki tipped her head back, delighting in the pain of his teeth.

"God I want you. And you know how much, don't you?" he muttered huskily.

"Yes." One syllable was all she could get out.

"I'd like to take you now, on the back seat of the damn car. You know that too, don't you?"

She tensed. "Michael—please—not—"

"Relax," he interrupted. "I want it to be perfect for you. I'll wait. But only until we get home, Nikki. You understand that?"

And he resumed his assault on her senses, teasing her as he had asked her to tease him, playing with her mouth, until she moaned, "Anything you say, Michael." Only then was she rewarded with the devouring, drugging kisses of moments before.

When he sought to carefully disentangle himself, she at first clutched his shirt to force him close again. She was too aroused to realize that he had something to say to her. He shook her gently, until she gazed anxiously up at him, her eyes half-lidded and lips soft.

"What's—what's wrong, Michael?"

"Nikki. Before this goes any further—I want you to understand—" He paused, as if he were having difficulty finding the proper words. "I find you incredibly appealing. That must be obvious. Maybe it's your innocence—I really don't know." He grinned crookedly. "Right now all I want to think about is what I'm going to do to you when I get you alone in bed." Then he abruptly sobered. "But I want to be fair. I want the arrangements spelled out. You're beautiful, and sweet, and I care for you. I want you to live with me—for as long as it suits us both. I wouldn't like to see you hurt because you expect love and marriage. That isn't in the cards."

Nikki felt so nauseated that she clapped her hand over her mouth to keep from retching. Oh you fool, you're so damn *stupid!* she cursed herself. You knew how he felt. Her stomach began to ache as if a mule had kicked her.

She forced herself to remember that she had some pride. She straightened out of her hunched position and tried to sound sophisticated. "I'm not a child, Michael. What—whatever made you think I needed a wedding ring?" She emitted a breathless, high-pitched laugh. "Maybe we should sign a contract. I'm getting an incredible bargain. By the time you get through teaching me—teaching me—" Absolutely nothing more would come out.

Michael raked his hair with his right hand. "Nikki—honey—I—Henry, look out!"

Henry Merola saw the truck at the same moment as he heard Michael's shout. By then the driver had almost careened into the Cadillac. Henry hit the horn and swerved; there was a screech of brakes as the truck driver tried too late to stop, skidded on the wet pavement, and hit the black Cadillac broadside toward the rear. Michael was thrown against the door of the car and Nikki was crushed between the hood of the truck and the man next to her. The last thing she was aware of was a stabbing pain in her arm, and her head crashing into the door next to Michael.

Chapter 8

Nikki opened her eyes, feeling thick-tongued and so tired that even speaking was an effort. She tried to remember where she was, but nothing came to mind. She examined her body. Her left arm was in a fingertip-to-elbow cast; a thin tube led from a hanging bottle into the front of her right wrist. It was filled with red fluid. A figure in white stood nearby.

She urgently sought reassurance. "Am I all right?" she asked, her speech slurred.

The woman nodded. "Yes. You're going to be just fine." Relieved, Nikki shut her eyes.

When she opened them for the second time, she was in a tranquil, powder-blue room. Her whole body ached.

"Nikki? Thank God!" A man was hovering by her bedside, looking extremely distraught. One side of his face was full of bruises and cuts.

She looked at him, wondering why he was spinning around. "Where am I? Who are you?" she moaned, but fell asleep before he could answer.

She slept until very early the next morning. When she woke up, a young nurse was with her.

"You're awake! Good. You've been moaning a bit. Do you want something for pain? The doctor left orders."

"Umm. Yes." Nikki still felt groggy. "I'm so tired."

The nurse took her temperature and blood pressure, then she picked up a hypodermic needle and expertly injected it. "You're tired because you took quite a beating in the accident, and also from the anesthesia. It was such delicate surgery and you were thrashing around. They had to put you out. I'm afraid this stuff won't help. You rest. I'll be right back."

She reappeared a few minutes later, Michael Cragun beside her. Nikki had never seen him look so pale. He approached the bed, reached out to stroke her hair, checked himself. He ran his fingers through his own uncombed hair instead. Then he pulled over a chair to sit down close by her. "Nikki," he asked tenderly, "do you know who I am?"

"The tooth fairy," she mumbled. "My mother—did you speak to her?"

"Yes. I told her you'd call when you were feeling up to it. My name," he persisted, "what is it?"

"Don't you know?" she asked with weak sarcasm.

"Do you?" he countered.

"Sure." As if she could forget! "Michael Lowell Cragun. Tycoon, lover, skunk." She noted with satisfaction that his face had reddened, then remembered seeing him earlier that morning and wondering at how scratched up he was. "You're all messed up," she stated undiplomatically.

"I got off easy. Just some cuts and abrasions, mostly from the flying glass. Thank God Cadillacs are well-built, or you—" His voice trailed off, as if he couldn't bear to think of what could have happened in a smaller car.

It occurred to Nikki that her own face might be bruised or cut. She touched her cheek with her right hand. "My face?"

"Beautiful, as always. Wait. I'll get a mirror." Michael went off to search the private bathroom. He emerged carrying a hand-held mirror, and held it up so that Nikki could inspect her appearance. She saw with relief that apart from a large bruised lump on her forehead, her face was undamaged.

"The rest of me?"

"A badly fractured wrist, a bruised body, and mild concussion. Your arm—it was badly cut by glass from the window. There are fifteen stitches in it. You lost a lot of blood. The plastic surgeon who stitched you up told me she did a beautiful job. But there will be scars—" He seemed consumed by guilt.

"It doesn't matter. What about Henry?"

"Not a scratch. The truck hit us toward the rear. He managed to steer away at the last minute. It helped."

Suddenly Nikki felt woozy and she giggled. "I think I'm drunk."

"It's just the shot. Go back to sleep now." He bent down to kiss her forehead, but her eyes were already shut. "I'll see you later, darling."

Nikki woke up again in mid-morning to the sight of sunshine streaming into the cheerful room. With its modern chests and print curtains it looked very little like a hospital room. Michael's college roommate Craig Landesman strolled in a few minutes later wearing a creased white hospital jacket, a stethoscope slung carelessly around his neck.

"Hi there," he greeted her. "I hear you got a bit too friendly with a very large truck. How are you feeling?"

"A little dizzy still," Nikki told him honestly. "And in pain. But I don't like to be shot full of drugs. It makes me feel like some sort of junkie."

"Don't worry. You've got an excellent doctor and you can trust him," the pediatrician assured her. He sat down by her bedside, and a suggestion of a smirk stole over his face. "I wasn't here yesterday evening when you were brought in, but Merry happened to be down in the Emergency Room. She's a resident here. If you can imagine Mike hysterical, I hear he was pretty close last night. Shows you what love'll do. He dragged Hal Worsley, the orthopedic surgeon, out of a dinner party. Hal called Barb Russo, the plastic surgeon who stitched up your arm. I'm told both of them did a first-rate job."

Nikki nodded. "I'm grateful. What about you—do you work here all the time, Craig?"

"I'm in private practice in Manhattan. I send almost all my hospital patients here—you're in Hudson Hospital, in case no one's bothered to tell you. What were you doing this far uptown?"

"I don't know." Nikki sounded puzzled; Craig's question was a good one. "Just driving around, I guess." A thought struck her. "I never asked about the truck driver. Was he okay?"

"Treated and released. There's a question of whether or not to press further charges. Frankly, Nikki, Mike's out for blood. Merry told me that if he hadn't been so worried about you, he might have creamed the guy."

Nikki considered that such a display of emotion from the usually controlled Michael Cragun would have been totally out of character. Merry was obviously prone to exaggeration. She asked, "What—what will happen to him?"

"I have no idea; I'm a doctor, not a district attorney," Craig answered wryly. "I just want to convince Michael to let the proper people handle it. With his influence he could probably wreck the guy's life. I'm sure he feels terrible already, and luckily no one was permanently injured."

"Are Michael and Henry still here?"

"This is beginning to feel like a game of twenty questions," Craig teased lightly. "But the answer is

yes—and no. Henry went home last night as soon as
you were out of surgery. Mike insisted on staying. He
slept in a hospital room down the hall and in his usual
arrogant fashion ordered the nurse to get him the
minute you woke up." He grinned at Nikki. "You
know, it took me three solid years of rooming with
Mike at Yale to finally learn when to tell him to go to
hell! Anyhow, I talked to him this morning, before he
saw you. He was convinced you had amnesia."

"He kept asking me who he was." Nikki's stomach
emitted a low growl, and she managed a smile. "I'm
hungry."

"Good sign. I'll rustle up a nurse to take care of you.
See you later, Nikki." He nodded as she thanked him
for coming to visit, and strode briskly out of the room.

She was just finishing her lunch when a bushy-
eyebrowed doctor poked his head into the room. Nikki
raised *her* eyebrows questioningly, and he came in,
introducing himself as Hal Worsley, the orthopedist
who had operated on her wrist.

"It was really messed up in there," he explained.
"Very challenging to patch up. I enjoyed it. I cleaned
you up on the inside and Barb Russo took care of the
outside."

"I'm glad I wasn't a boring case," Nikki joked,
"seeing that you were hauled out of a dinner party. I'm
very grateful to you—"

"Thank your boy friend. He's like a human bulldozer
when he wants something." His tone told Nikki he was
teasing, but she of all people knew that there was more
than a little truth to his observation! "Actually," Dr.
Worsley continued, "my mother-in-law's dinner parties
are more boring than the simplest fracture. I only go to
placate my wife. You rescued me from a very determined elderly matron who was reciting a list of

symptoms to me as if she were reading from her grocery list. I didn't have the heart to tell her that I only do hand and arm surgery."

"Occupational hazard, doctor?" Nikki asked drily. But her attempt at amusing repartee was overshadowed by the look of pain in her eyes, and her doctor hastily changed the subject.

"I think we'd better talk about you, Nikki. Mike wants to spirit you home today, but I'm going to have to insist that you stay until tomorrow afternoon. Frankly, although you escaped with only a shattered wrist, some stitches and a concussion, your body took quite a beating. I'm amazed that there were no broken ribs or internal injuries. I guess you know Mike managed to protect most of you with his own body—otherwise it might have been a lot worse. I'm going to leave orders for painkillers as you need them—say every four hours."

"I don't like to take that stuff—" Nikki timidly objected.

"Fine. Do without it as much as possible, but don't be Spartan about it. You're pretty battered, and a few days of drugs won't hurt you. I did speak to a neurologist on the staff about you, by the way. But if you'd like another opinion, I can get an internist friend of mine up here or anyone else you choose."

"Thanks." Nikki smiled, "I hear I've already got the best."

"I'm very susceptible to flattery," he responded, "especially from beautiful young patients. I'll be back to check on you tomorrow. We can set up a schedule of office visits then." His voice assumed a confidential air. "Keep it quiet though, or everybody will expect the same red-carpet treatment. Usually my patients make appointments through my office, and they suit *my* convenience. But Mike's going to want to come with you, and he's even busier than I am. Besides, nothing's too good—"

". . . for a friend of Michael Cragun's," Nikki finished. "Now where have I heard that before?"

"Any number of places, I imagine. Now get some rest, young lady," he told her in his most authoritative manner.

Nikki had accepted some pills proffered by the nurse, and was watching an actor plug his latest film on a talk show when Michael came in. Dr. Worsley's remark that he had saved her from more serious injury had put him very much back into Nikki's good graces. Her hurt and resentment had been relegated to a distant corner of her mind. Beyond that, she had simply repressed thoughts of Michael or the accident.

Michael picked up the television remote control switch from the bed and clicked off the talk show. "How are you feeling?" he asked Nikki.

She was too much smitten by his self-sacrificing nobility to object to his action. "A bit better. And Michael—Dr. Worsley told me what you did. That you threw your body over me in the car. I can't thank you enough for—"

"Don't!" he snapped. Then he sat down, and to Nikki's bewilderment she saw weariness and guilt chase over his face.

"This is going to sound like ·a *True Confessions* article," he said.

"Then skip it," Nikki put in quickly, disconcerted by his self-reproachful behavior.

"No." He gazed across at her. "Just don't start thanking me. If I hadn't told you to take your coat off, it probably would have protected your arm. I—I saw the truck a moment before it hit us, and managed to shelter most of you. But your arm—" He shook his head.

"That's a ridiculous thing to feel guilty about," Nikki scolded. "The coat was soaked, and—"

"You just don't understand!" he exploded. "Why in hell do you think we were driving all over Manhattan? I told Henry not to go home right away. He gave me one of his patented stares of disapproval, but he's learned to follow orders. You set out to needle me Thursday night, and by God you succeeded. If it hadn't been one o'clock in the morning when Jack and Shaun left, we would have had it out *then*. As it is—damn it, Nikki, I set you up. I got you in that car and I decided I wasn't letting you out until you were begging for it. Of all the stupid things I've done—oh, the hell with it!"

Michael got up so abruptly that he knocked the chair over. He impatiently righted it, then stalked off into the bathroom. Nikki heard the sound of the tap being turned on and heard him gulping down a glass of water. She lay motionless, slightly stunned by his admission, but nonetheless feeling that his guilt was all out of proportion to the alleged crime. True, his plans were cold, even callous, but she had prodded him shamelessly all week, and been so rude and provocative Thursday that she was probably lucky to survive intact until Friday morning.

When Michael came back into her room, he was calmer but his face was filled with blatant self-disgust. "You called me a skunk before," he said bleakly. "You were being kind."

Nikki had decided that Michael was a hero and was not about to change her mind. His self-recrimination engendered only her sympathy. The fact that he was conscience-stricken over so minor a matter as an attempted seduction only made him rise in her estimation. She was determined to make him realize that self-flagellation would accomplish nothing.

"Michael," she said sweetly, "you can't blame yourself, you weren't driving the truck—"

"When I get my hands on that damn driver, I'll see that he doesn't get near another truck in this city for as long as he lives!" was the snarling reply.

And then Nikki lost patience with his negative behavior. The imperiousness of his last statement caused something to snap inside of her. "Stop it!" she all but screamed. "You think you can tell the whole world what to do, but you can't, Michael Cragun. Ever since I answered that—that insane advertisement of yours, you've been pushing me around. One minute it's with threats, the next with that charm of yours that you turn on and off like a faucet. Now all of a sudden you're swimming in self-pity and I'm dumb enough to lie here and join you in feeling sorry for poor Michael. Well, forget it! You're going to listen to someone else for a change. First! You are going to let the police and that driver's employer deal with him. Second! You're going to stop this self-indulgent wallowing of yours. I'll live! And third! You're no doubt spending a fortune on doctors—you'll let them take care of both of us!"

Having delivered herself of this heated tirade, Nikki contented herself with glowering at him, a look of utter exasperation on her face.

For a moment Michael scowled back at her, obviously furious that a slip of a twenty-three-year-old girl had dared to speak to him in such a disrespectful manner. His affronted dignity was so comical that Nikki couldn't prevent a giggle from escaping. She clapped her hand over her mouth to stifle it.

Michael ruefully shook his head and smiled crookedly down at her. He dropped lazily back into the chair beside her bed. "I haven't been chewed out like that since I was twelve years old," he told her, chuckling. "Even my father hasn't had the guts. You have any more orders, ma'am?"

Nikki's eyes sparkled. "No, but since I have you at my mercy, I'm going to get some answers. That attack of conscience of yours in the car—was that part of your award-winning performance?"

The humor drained from his face. He was silent for so long that Nikki figured he wasn't going to answer at all. At last he shrugged, "You want the truth? I don't really know. All the time I was saying it, I felt myself standing two feet away, admiring my acting ability." His voice became husky and he took her right hand. "Nikki, I was wrong about one thing I said in the car. I told you I wasn't interested in love and marriage. You were so hurt—I'll never forget the look on your face. Honey, when I saw you lying there unconscious, I knew—"

"Michael," Nikki interrupted wearily, "haven't you done enough?" She didn't believe a word he had said. He had been irrational during this entire conversation. "I don't want to hear any declarations of love prompted by this stupid guilt of yours. I don't hate you, okay? Now let's just drop it!" She winced as a sharp pain in her arm joined the steady throbbing of her head.

Michael's face hardened. Nikki knew that look. It meant he was bent on having his way. "Yes, all right. For the moment, Nikki. I don't want you to upset yourself. You've given me a slew of orders, and now I'm giving you a few. After you leave the hospital, you're coming home with me. You're going to let the Merolas—and me—look after you until your arm is out of that cast and the sublease on your apartment runs out. Then if you still want to leave, you can leave. Agreed?"

As if I have any place else to go! Nikki thought. But leaving Michael was the last thing in the world she wanted to do. His insistence on taking care of her was only a sop to his guilty conscience, but it meant a great deal more to her. A small part of her dared to hope that she had some future with Michael, but she deliberately quashed any such dream. Less than a day ago, when he had been in a considerably less agitated mood, he had made it clear that he didn't love her and would never marry her. She would have to accept that.

Nevertheless, she loved *him* and wanted to be with him. Time enough to worry about the pain of separation in six weeks.

She forced a noncommittal look onto her face, and cocked her head. "Sure. Why not? I've gotten spoiled by a life of luxury." Her face twitched as pain shot through her right temple. She pressed her right hand to her head and kneaded her scalp.

"Here, let me," Michael said quietly. He reached over to gently massage her head and neck until she closed her eyes, gave a contented moan, and dropped off to sleep.

Nikki began to feel as though her room was in the midst of Grand Central Station when the Merolas dropped by to visit at dinner time. Suzanna, highly mistrustful of hospital fare, had brought along several thermos bottles of food—the inevitable chicken soup, a bland version of Moo Goo Gai Pan and some cold tapioca pudding. She announced to Nikki that she would look after her "better than all the boss's fancy doctors." After comparing Suzanna's dinner with the hospital's ground meat, Nikki enthusiastically concurred.

Henry was unusually subdued, his thoughts still haunted by visions of what might have happened. He clearly considered himself responsible, but Nikki would hear none of it. The show of affection from both Suzanna and Henry touched Nikki deeply. It also made her feel safer because she knew that Michael's contrition of today could easily be forgotten by next week.

Nikki felt well enough that evening to phone her mother and reassure her about the extent of her injuries. Because Dr. Worsley was concerned about Nikki's concussion, he instructed her to stay off her feet and rest as much as possible for the next week or so. She explained this to her mother, adding that she would visit as soon as the doctor gave his permission. Mrs. Warren replied, "Don't worry, darling. Stephen's been spending—too much of his time with me. I've been working with the therapists. You take care of yourself."

"And let Stephen Rowland take care of you?" Nikki couldn't resist asking.

"Yes. And I rather like it, Nikki," her mother had answered.

Nikki was discharged from the hospital late the next afternoon. She insisted on walking out; hospital policy dictated she be seated in a wheel chair. Whereupon Michael picked her up, dumped her unceremoniously into the chair, and pushed it through the hall into the elevator. In the lobby he picked her up and carried her into the car. At his apartment he repeated the exercise, and Nikki, delighted by the experience of being held in his arms, made no objection.

She stayed home for the next week, shamelessly enjoying her brief reign as the tsarina of Fifth Avenue. Suzanna and Henry were so eager to spoil her that it took all of Nikki's will power to discourage them. But some devilish streak in her makeup prodded her to find out just how far she could go with Michael before he rebelled. His humble desire to wait on her was so out of character that she decided he was begging for trouble.

Every time he asked if he could get her anything she managed to manufacture some item—always from downstairs if not one of the nearby stores—that she just simply could not live without.

In the middle of the week she started going downstairs for breakfast, then returning upstairs to read or watch television. Nonetheless, on Saturday morning, Michael appeared with breakfast in bed—a cheese omelet, orange juice, toast and hot chocolate.

It occurred to Nikki that he was rather smugly pleased with himself as he set the tray down. Tongue firmly in cheek, she told him, "You've surprised me this week, Michael. I never pictured you as having so much patience."

He was egotistical enough to consider it a compliment. "I know it's hard to be in bed," he said solemnly. "I hope I've made it less boring for you."

Nikki studied the contents of the tray as Michael sprawled into the armchair to keep her company while she ate. "Michael," she said innocently, "I'm not in the mood for orange marmalade. Could you get me some strawberry jam instead? Please?"

"Sure. I'll be right back," he said willingly. He disappeared from the room; Nikki heard him dash down the steps two or three at a time. Soon he was back, jam in hand. Nikki waited until he was again relaxing in the chair before dramatically picking up her orange juice, sipping it, and regarding it with a screwed-up face.

She said to herself, "Gee, I usually love orange juice. I can't think why it tastes so bad this morning." She turned to Michael. "Is there any other kind of juice around?"

This time he did not jump up from the chair to satisfy her demands. "What kind did you want?" he asked, and Nikki suppressed a smile at the testy note in his voice.

Nikki pretended to ponder her options. "Umm. Grapefruit, maybe."

"Maybe?" Michael lifted his eyebrows.

Nikki looked more certain. "Grapefruit, definitely."

He shot her a quizzical look, nodded slowly. "Right." Nikki noted that he took his time with this second request.

As he came in, she was munching on a piece of toast. She had deliberately left off the jam, and judging by the obvious manner in which Michael looked at the toast, then at the jam, and then at her, this had not escaped him.

"Gosh I'm thirsty," Nikki said. In response Michael held out the juice. She shook her head. "No. For water, I mean. Could you get me some ice water from the fridge? Please, Michael?"

He stood by the bed, legs encased in tight-fitting blue jeans, a white tee shirt covering his chest, feet bare. He reached up a hand to rub his unshaven jaw, a sardonic look in his eyes. He appeared to be trying to determine if Nikki could actually be teasing him, or if she were simply a difficult patient.

Nikki made up her mind that since Michael seemed disinclined to run downstairs to fetch her the water, she would have to encourage him to be more solicitous. She looked up at him appealingly. "I'm being too demanding," she said contritely. "Never mind the water."

"No, no, it's okay. I'll get it." He walked to the door, looked back at Nikki and asked, "Are you *sure* you don't want anything else?"

"Yes. Absolutely."

When Nikki heard him start down the steps, she could contain her laughter no longer. She barely managed to sober up by the time he returned with the ice water. She plunged the knife a little deeper. "Will you get me the *Vogue* magazine from the top of the television set?" she asked.

Michael stared at her, not moving, and Nikki sensed the moment for total outrageousness had arrived. "You know, Michael, as soon as I'm better, I'm going to Tiffany's to get you a gift."

He cocked an arrogant eyebrow at her. "Oh?"

"Mm. I'm going to buy you one of those gold choker collars—the ones that look like slave collars."

She clenched her teeth to keep from laughing at his incredulous look. Hands on hips, he accused, "You're enjoying this, aren't you?" It was more a statement than a question, and he didn't bother waiting for her reply. "You've been putting me on all week, trying to see how far you could go with that wide-eyed routine of yours, and I just realized it."

Nikki promptly succumbed to hysterics; she laughed so hard tears rolled down her cheeks. "Lady," Michael said dangerously, ignoring her laughter, "you're lucky your arm is still in that cast. Or right now you'd be lying over my knees and my hand would be connecting very forcefully with your well-padded bottom."

"You wouldn't," she challenged.

"Not only would I, but I'd thoroughly enjoy it." His eyes glittered. "Not too hard, of course," he told her wickedly, "and as for what would come afterward—" His eyes traveled over her body, which was outlined under the light covers, and came to rest on her mouth. Then abruptly his mood changed. "Finish your breakfast, Nikki," he told her stonily, and stalked out.

She did so, a satisfied smirk plastered on her face. She had not missed Michael's thorough appraisal of her considerable physical charms and knew that she ran the

risk of retaliation when she teased him. Yet it was only
human to give in to the temptation to do so. In the
weeks since she had first become entangled with
Michael Cragun, he had shown little inclination toward
virtue. Now he was determined to be a gentleman, a
big brother, and Nikki considered it a great game to
provoke him into acting the opposite.

Erika visited Nikki the following Sunday, and they
spent the afternoon talking—about everything except
Nikki's relationship with Michael Cragun.

Nikki told Erika that the next morning couldn't
arrive soon enough—she was aching to get back to
work. Five days of television and forced inactivity had
left her restless, edgy, and ready to tackle even the
most boring of routine paper work. The stitches had
been snipped out of her upper arm on Friday, leaving a
jagged scar. Nonetheless Dr. Russo was pleased, and
assured Nikki that the cut was healing well.

Dr. Worsley removed Nikki's cast the same day,
replacing it with a new one. He pronounced himself
satisfied with her progress, and since she had experienced no further headaches or dizziness, approved her
return to work. Amused at Michael's frown of disapproval, he merely cautioned her not to overexert
herself.

Life resumed its normal course for Nikki, if limou-
sines, gourmet dining and elegant apartment living
could be considered normal. Much to Nikki's regret,
she had scant opportunity to goad Michael. Not only
did he work long hours, but he traveled constantly.
CAI was extending its tentacles to Europe and Michael
needed to confer with corporation executives in England, France and Germany. To Nikki these casual
transatlantic jaunts were ineffably glamorous. Michael

assured her that he saw the insides of jet planes, board rooms, hotels, and little else.

Michael did make it a point to be in New York on Fridays so he could accompany Nikki to the doctor. She accused him of being overprotective of her, but was secretly pleased by his attentiveness. She impishly sat close up against him in the back seat of the car during the twenty minute trip to and from Dr. Worsley's office. Unfortunately, he seemed oblivious to her nearness. She had no greater success in arousing either his lust or his annoyance on the longer rides to visit Pam Warren, who was transferred to the March Institute several weeks after Nikki broke her arm.

If Nikki had unconsciously hoped for some further declaration of love from Michael, she was in for a rude disappointment. He was solicitous, friendly and protective, treating her as he would a favorite young cousin or friend of the family. During what Nikki poutingly called his pit stops in New York, he set about providing entertainment with the same efficiency with which he ran his business. The days were so full that Nikki felt positively programmed. Michael took her walking and picnicking in Central Park, where Nikki deliberately sprawled back on their blanket, pretending to relish the balmy weather but far more interested in Michael's reaction to her attempts at seductiveness. To her chagrin, he had no reaction at all.

They went shopping at the most exclusive department stores in Manhattan in order to find loose-sleeved clothing that would fit over Nikki's cast. Michael expressed no particular desire to inspect Nikki's appearance in the expensive clothing he would shortly pay for. He left her to the mercies of anxious to please saleswomen and sat reading reports and making notes while she shopped.

He escorted her to plays, operas and concerts, invariably obtaining seats in the first ten rows and inevitably running into friends and acquaintances. Nikki was taken aback by the number of people who made comments such as, "So you're Nikki," or "Nice to finally meet you" or "You're as beautiful as everyone says." She could only smile and mumble some appropriate comment, wondering why Michael's arm was always draped over her shoulder in public when in private he was so resolutely impersonal.

Nikki generally read only the New York *Sun* in the mornings. Not so Michael, who was an avid newspaper reader, glancing through at least five papers daily. Nikki came down for breakfast one Monday morning to find a copy of the *Sun's* competition open on the table, an item in the gossip column circled: "Tempestuous diva Elianna Dunne is singing off-key these days," it read, "since her great and good friend, CAI boss Michael Cragun, turned his attentions to an ethereal brunette with a broken wrist. Rumor has it the two were introduced by the lady's boss, Cragun's pal Peter Delavan, the New York *Sun* publisher. Why did you dump La Dunne, Mike? Did you find out she was two-timing you with her conductor?"

Michael had scrawled across the margin, "Always the last to know! Thought you would enjoy a laugh at my expense, ethereal brunette." He didn't, Nikki reflected, seem overly broken up about it.

In the middle of May, Michael's eleven-year-old niece came to spend the weekend with her uncle. The slender daughter of Melanie and Brandt Newman clearly adored her Uncle Michael, while he derived amused pleasure from squiring the enthusiastic young girl around town. Nikki noticed a bit sadly that Michael treated his niece in very much the same manner as he treated her, although for the lucky Lauren there was abundant physical affection in addition to avuncular indulgence.

Laurie had started taking dancing lessons when she was six years old, and her love for ballet had never waned. Other preteenagers had posters of rock stars and actors in their rooms; Lauren told Nikki that autographed pictures of Gelsey Kirkland and Natalia Makarova hung over *her* bed. Other eleven-year-old females stood before their mirrors anxiously inspecting their bodies for the approaching curves of womanhood. The elfin Laurie, all boyish slimness and angular grace, dreaded the possibility that her dancer's body would be ruined by maturity. Nikki solemnly assured her that if heredity had its way, she would grow into a slightly taller copy of her idol Gelsey Kirkland. Even after two pregnancies, Melanie Cragun Newman retained the small-busted, narrow-hipped body of a model or dancer.

Nikki considered it quite an act of sacrifice that Michael was willing to sit through both Friday evening and Saturday afternoon ballets with his niece. Much to her amazement, she learned that he was a knowledgeable balletomane. Friday's performance had starred Fernando Bujones and Cynthia Gregory and Lauren was awash with gushing admiration. But some of the ballets in the matinee featured younger, less spectacular dancers, and uncle and niece lambasted their efforts even more harshly than the sophisticated New York audience, which sat on its hands throughout much of the afternoon.

After Saturday's performance, they went to a nearby restaurant for seafood. Michael relished the fact that Nikki had to ask for his assistance in cracking her

lobster. He had not offered any help, but lazily watched
her struggle one-handedly with the crustacean, until
she was finally forced to seek his aid. He teasingly
offered to feed it to her as well as crack it, and she
fervently wished that she had ordered shrimp or filet of
sole instead. Michael's gentle mockery was harder to
endure than his big-brotherly indulgence; it was so
tempting to imagine that he meant it differently. But he
treated Laurie in exactly the same fashion.

Next to ballet, Laurie loved animals more than
anything in the world, even, Michael said with pretended offense, himself. Laurie requested that they spend
Sunday at the zoo; Michael countered that his handball
game was sacrosanct. They compromised on sports in
the morning and animals in the afternoon. Returning
two hours later, he requisitioned Suzanna for a picnic
lunch, and marched Nikki and Lauren out to the car.
He hustled them both into the front seat, Nikki beside
him, his arm around her shoulder. Lauren gave Nikki
the type of look that only an eleven-year-old who
knows everything can carry off.

They sat in one of the zoo's restaurants to eat their
picnic lunch, then took a tram ride around the park.
Afterward they set out on foot for a closer look at the
animals, most of which were displayed in natural,
outdoor settings. Laurie headed straight for The World
of Darkness, a perennial favorite with visitors to the
Bronx Zoo. At that exhibit, day and night were
reversed for the animals by turning bright lights on
during the nighttime and using dim, infrared lighting in
the daytime. Thus, the normally nocturnal creatures
scurried around during the hours when the zoo was
open, much to the delight of the many onlookers.

One dignified looking owl was perched on the limb of
his tree, but quickly ducked out of sight as they passed.

When they stepped out of his line of vision, the owl carefully peered around his tree once more to find them. They appeared; he retreated. They played this game with the owl for several minutes, and then walked on to watch the bats soaring back and forth in their large flyway.

Laurie and Michael seemed to have no dearth of energy, but after four hours of lions, deer, elephants, birds and bison, with only a brief break for drinks, Nikki could hardly stand up. Not wanting to spoil Lauren's afternoon, she said nothing, but ultimately stumbled and fell, only to be snatched up by Michael just as her cast was about to connect with the concrete. For a moment, his face was thunderous, as if he were ready to deliver a sound tongue-lashing. But he contented himself with scolding her as if she were a naughty child for not saying she was tired, and summarily hauled his two female companions back to the car.

They headed away from the city toward the home of Lauren's parents in Dutchess County, two hours north of the zoo. In his usual commanding fashion Michael ordered Lauren into the back seat to rest and directed Nikki to lie down in front. She soon fell asleep, her head on his thigh, her legs curled up on the seat.

As soon as they walked in the door of the Newmans' converted farmhouse, Lauren's eight-year-old twin brothers assaulted their uncle, roughhousing with him until he begged for mercy.

"I have an idea!" Laurie piped up. "Tell Uncle Michael and Nikki they have to come back, and babysit us."

"You want to run that one by me again, sweetie?" Michael drawled.

"Mom and Dad are going to an antique show in Massachusetts in two weeks, and they were going to get

Mrs. Grey, but I'd rather have *you*, Uncle Michael, and of course Nikki too, because otherwise I'm the only girl," Lauren said without stopping for breath.

Melanie's "Really, Lauren!" was interrupted by a clamor of approval for their sister's plans from Mark and Jeffrey. The twins complained that Mrs. Grey was old and mean and no fun at all, although their mother retorted that from the way they hugged her good-bye each afternoon, one would never know they thought her such a dreadful witch. A flattered Uncle Michael speedily caved in to niece and nephews' unabating insistence, telling them, "I know when I'm licked. But I'll only do it if you can convince Nikki to come too."

"Michael, I'd love to spend a weekend up here. It's beautiful. But I wouldn't be any help at all. With this arm in a cast, I can't even cook a meal or wash a dish."

Her reluctance was futile, as the others cajoled and nagged and reassured her, ultimately overriding her every objection.

"That's settled then," Melanie said with satisfaction. She cocked her head toward her brother, the mocking gleam in her blue eyes reminding Nikki of the times Michael had looked at her like that. "Nikki can stay in with Lauren. I hear you're sleeping alone these days, Mike."

"Much to my regret, big sister," he laughed, "Nikki has some outdated notions about love and marriage."

Nikki cursed the blush that rose to her face. She would have liked to kick Brandt for chiming in with, "Then you'll just have to marry her, won't you?"

Michael answered laconically, "Exactly." Then he took Nikki's arm and led her out to the car. She sat as far away from him as she could, brooding about the scene that had just taken place. Once they were on their way she said vexedly, "Your whole family assumes we're going to get married. Why don't you set them straight?"

"Don't you want to marry me?" Michael asked coolly. "You've been teasing the hell out of me for nearly a month. What's the idea? Seduction followed by a proposal?"

Nikki had no idea what to say. His analysis was embarrassingly close to the truth.

"I've compromised your spotless reputation. Items in gossip columns, everybody knowing we live together. I have to make an honest woman of you, don't I?" he mocked.

She muttered, "Don't be ridiculous."

"Ridden by guilt, irrational, ridiculous. You have a high opinion of me, Nikki," Michael drawled.

"You know what I mean. You don't love me, you don't even want me. You treat me like a little sister. You don't even notice when I—when I—"

"Yes?" He was smiling over at her.

"When I provoke you," she blurted out.

"Don't I?"

"No."

He sighed. "Okay, have it your own way, Miss Warren. I don't love you. I don't want to take you to bed with me. I'm not going to marry you. But I'm damn well warning you, the next time you drape yourself all the hell over me, I'm carrying you into the nearest bedroom. Cast or no cast, I've taken all I'm going to take."

Nikki wanted to ask: Do you love me? But no words came out. She was honest enough to admit that even a positive response wouldn't satisfy her. Michael was too far out of her league for her to believe that he was in love with her. It occurred to her that perhaps he did intend to marry her—but for reasons other than undying adoration. To satisfy his family, to have children, because he felt he owed her a wedding ring after turning her life upside down and indirectly causing her serious injury.

He had made it clear that he found her attractive; making love to her would hardly be an onerous chore. But she had no intention of agreeing to any such marriage of convenience. She had been bluntly warned not to push him any further. She would heed the warning.

Chapter 9

If I still speak to you after this, Nikki silently told Michael Cragun, I'm either totally besotted or nuts!

Michael had at first refused to believe that anything could possibly be wrong with him. He had been in Los Angeles, Dallas, Chicago and Florida since the previous Monday, and had returned to New York Saturday with a persistent cough that he attributed vaguely to allergies. It was spring, he claimed, all sorts of vicious pollens lurked in the air. He refused to skip his weekly game of handball on Sunday afternoon, growling at a hapless Suzanna who had had the temerity to suggest that he stay home to rest. To no one's bewilderment but his own, he arrived home drenched with perspiration and totally exhausted.

Once bitten, twice shy. Suzanna had regarded her boss with tight-lipped reproach. But Nikki had had the unpardonable gall to express concern over his condition. Her reward: a violent "I just need a cool shower; stop your damn nagging!"

Michael overslept Monday morning, not waking up until 7:45. As a result he had ridden to his office along with Nikki. She had been too cowed by his foul temper and sharp tongue to point out that breakfast should consist of more than a hastily swallowed cup of black coffee. Even Suzanna, who was insulted on principle when anyone refused to eat her cooking, had reined in her reproofs.

Michael had helped Nikki out of the car in front of the *Sun* building, and had been thoroughly annoyed by her alarmed concern as she felt his hot, moist hand. He snarled, "Just stow it, Nikki," and all but slammed the door in her face.

She was not surprised when on the way home, Henry informed her, between sneezes, that "the boss" was already home—in bed.

Michael slept through until the next morning. As Nikki dressed she could hear him all the way from her room, lashing out at Suzanna, who had committed the inexcusable sin of suggesting that a doctor be called. When she peeked in to say a solicitous good morning, his expression was so forbidding that she hastily shut the door again.

Mrs. Warren had been making wonderful progress since her move to the March Institute. Nikki decided to make a brief visit Tuesday after work, and asked Henry to wait until she was ready to leave. She found Pam Warren in her room, and the two of them walked slowly down the hall to a brightly-colored lounge to play several games of casino together. As Nikki was about to leave, Stephen Rowland appeared, kissed her mother hello, and produced a small velvet case from his inside jacket pocket. He calmly removed the diamond from inside, took Pam Warren's left hand to divest her ring finger of its wide gold band, and slipped on the diamond engagement ring he had bought her.

Then he said coolly, "You don't have the option of saying no. I'm telling you in front of Nikki so there'll be no backing out."

Nikki had never seen her mother glow with such happiness as when she murmured, "Whatever gave you the idea that I would think of refusing?"

There were hugs and kisses all around, and as Nikki said good-bye to her future stepfather, he told her with a smile, "Love and marriage is the best therapy for all kinds of ailments. Including strokes—and broken wrists."

Nikki arrived back at the apartment to find the usually unflappable Suzanna as irritable as her impossible boss. She was muttering to herself in Italian, and when Henry walked in after parking the car, she took one look at him and told him in English, "One patient is bad enough around here. If you don't get into bed—" A tirade in Italian ensued.

Henry placatingly hurried off, to Suzanna's exasperated, "Men! Such children!"

And so Nikki cajoled Suzanna into letting her help look after Michael during the evenings, because after nine hours of her boss—and now, it would seem, her husband—Suzanna's patience was worn to a frazzle. Worst of all, the fractious patient refused to get the sleep he so obviously needed. He forced himself to stay awake to work and was all the crankier for constant tiredness.

Smiles, pleases and thank-yous, which were never

Michael's strong point, had disappeared entirely from his repertoire. For two evenings he barked out orders at Nikki's docile head: get him something else to eat, he needed new ice water, the heat was too high, why the hell was it so sweltering in the bedroom, how long was it going to take Nikki to get Mr. Chessman on the phone in California because he didn't have all year, where the devil was the report he'd been reading earlier in the day. At first Nikki suspected he was taking his revenge for her own demanding behavior of several weeks before, but she soon concluded that his sense of humor had vanished. Her stoic patience soon vanished along with it.

She said angrily, "I'm limiting the amount of time I spend with my mother to cater to your unreasonable demands. I'm not your slave, and I won't be treated like I am!" By now Nikki was sputtering out the words, almost too enraged to be coherent. "I work all day, drag back and forth on that rotten bus, and I don't need to come home to a spoiled, rude autocrat who never would have gotten this sick if he'd realized that he can't push himself as if he were still nineteen years old!"

"Good! Terrific! Get out then! You hover over me like some kind of frustrated Florence Nightingale! Who needs you!" he stormed back, and for good measure added, "And take a cab to work, you little idiot!"

To her disgrace, Nikki burst into tears and fled the room. Nevertheless, when she was greeted on Thursday by an exhausted Suzanna, she glumly resigned herself to another evening of playing whipping girl.

But Michael was seethingly polite. Every order was preceded by "If it isn't too much trouble for you" or "Would you mind terribly" or "I'd be everlastingly grateful to you if." Although the temperature in the apartment was 72°, Michael insisted that the room was freezing. Nikki turned up the thermostat to 78°. There might be an energy shortage in the nation, but Nikki Warren was suffering a personal energy shortage and had no fuel left to argue with him.

Hot air poured out of the vents. Stifling, Nikki changed into a scoop-necked, sleeveless pink shirt and white and pink shorts. She sat down to watch television. An impatient voice carried down the hall.

"Nikki!" Sighing, she got up to see what he wanted. It was at that point that she had observed to herself that she was either madly in love or crazy.

Michael greeted her with a frigid, "Thank you so much for coming. I appreciate it more than I can say. What on earth have you got on?"

"I'm sure you didn't call me in here to discuss my clothing," a goaded Nikki said through clenched teeth. "What do you want?"

"I can't find the Pacific Timber contract. If it wouldn't be a terrible inconvenience—" Sarcasm dripped from his tongue.

Nikki, who felt like kicking him, regarded his bed with dismay. Papers and folders were scattered helter-skelter all over it. She let out a noisy breath and sat down cross-legged on the bed to methodically sort through the mess. She could feel Michael's eyes on her and fervently wished it weren't taking so long to locate what he wanted. Having only one good hand slowed her down considerably. Finally she found the contract sandwiched between the pages of a report, put it into the proper folder, and handed it to him.

"Thank you so much." Michael threw the folder on the floor, and said in his familiar drawl, "You really want to improve my mood? Then take off what little you're wearing and come make love to me."

Nikki refused to take him seriously. "In the shape

you're in, you'd probably collapse before anything could happen," she informed him witheringly.

For the first time in five days, he actually smiled. "Sad, but true," he lamented, and picked up the folder.

On Friday morning Nikki was disgruntled to hear Michael rummaging around downstairs at seven o'clock. He would wind up twice as sick if he refused to take care of himself. Nikki had decided to phone Melanie Newman later that day to suggest that the children come to the city for the weekend. She had avoided reminding Michael of their promise to babysit, reasoning that it could wait until the morning. Now she braced herself for the inevitable clash.

But when she threw on a robe and marched downstairs to confront Michael, she found that he looked almost healthy this morning. A slightly pale face was the only sign of his recent bout with the flu. He firmly told her not to start babbling at him, he was all right. He would see her at 4:30, and since he was going to pick her up directly from work, she should get her tail upstairs to pack her suitcase. He would put it into the trunk of the car before he left for work.

They listened to the first game of a twinight double-header on the radio as they drove through the crush of fellow escapees to the country. Nikki had half-expected an apology from Michael for his intolerable behavior toward her and indeed everyone, but he was mute on the subject. Perhaps he assumed that all flu patients were tantrum prone tyrants. In any event, Nikki felt she could survive this weekend only by feigning cheerful impersonality. She was hardly about to start demanding apologies.

Melanie popped their dinner into her microwave

oven as soon as she noticed her brother's automobile coming up the driveway. The minute Nikki and Michael walked in the door, she hastily reeled off a set of instructions, then produced a piece of paper on which the identical information had been written down. After kissing their children good-bye, she and Brandt effusively thanked Michael and Nikki and set out for Massachusetts.

After a few hours with Jeffrey and Mark, Nikki had to admit that she understood their precipitate departure. The twins, frantically excited by their uncle's visit, had been two little dynamos all evening. A game of Monopoly turned into a screaming match, with each twin accusing the other of cheating. Nikki finally had to put her foot down.

She firmly ushered the three children through showers, tooth-brushing and story-time. A cowardly Michael had taken refuge in his brother-in-law's study. When summoned to say good night, he had pretended to be reluctant to venture out to bestow kisses on his rambunctious nephews. As he went upstairs to their rooms, Nikki was amused to hear him mutter, ". . . and I *wanted* children?"

Later, the two adults sat on opposite sides of the living room, reading. Nikki relished the silence, which was punctuated only by such country noises as wild animals scurrying about in the woods hunting for food, and the wind rustling through the trees which surrounded the house. After a few minutes Michael strode over to his chair, where he had carelessly slung the jacket of his dark pin-striped suit. He removed a rectangular-shaped jewelry case from the inside pocket and handed it to Nikki.

"For you. For putting up with me," he explained brusquely.

Inside was a delicately filigreed diamond and emerald necklace with matching earrings. Nikki could feel her body growing hot. She tried to tactfully refuse the exquisite gift.

"Thank you, Michael, but my mother taught me never to accept presents like this from men like you," she said lightly. "It's quite a peace offering, though. You have beautiful taste." She blurted out a not so tactful afterthought: "Did you pick it out yourself?"

He threw her a mock scowl. "Yes, Miss Cynic, I picked it out myself! To match your very striking green eyes. Let me help you with the necklace." He sat down on the couch next to her, removed the necklace from its case, and fastened it around her neck.

Nikki forced his hand away before he could finish securing the intricate clasp. "Michael, I can't let you give me this. It's far too expensive. Besides, a simple apology for your abominable crankiness would have sufficed."

"I'm not very good at apologies," he admitted. "I *am* sorry, Nikki. I haven't been sick since I was in college and I couldn't stand feeling exhausted all the time. I freely admit that you should be canonized for putting up with me. Now stop arguing and let me put the necklace on you. I assure you my bank balance was scarcely dented." He finished tightening the clasp of the necklace, his hands cool and impersonal.

Nikki sat stiffly, wanting him to touch her, wanting him to pull the pins out of her hair, wanting him to kiss the back of her neck or slip a hand under her sheer, low-cut blouse. Instead, he handed her the earrings to clip on, and surveyed the result with detached admiration, Nikki sensed, for his own good taste.

"Umm. Suits you very well." He grinned. "You can thank me any time."

Stung by his unforgivable failure to try to make love to her, Nikki said peevishly, "I've already been warned about that. Tell me, did you buy trinkets like this for everyone else?"

"No—I *pay* them, and since I'm usually such an exemplary boss . . ." he joked.

"Really? As a one-time prospective employee, I wouldn't say so." Nikki tauntingly held up her broken wrist. "Look where my relationship with you got *me.*"

Michael uttered no stinging rejoinder to this, but simply sat motionless, staring bleakly ahead. Nikki was immediately ashamed of her below-the-belt sarcasm, and said contritely, "Oh, Michael! I'm sorry. That was a foul thing to say, and I didn't mean it at all. You know I don't really blame you for what happened. The jewelry—it's beautiful, but you must see that I can't accept it. It's—it's like some sort of bribe—to make up for a guilty conscience. Isn't it?"

"Is that really how you see it?" he questioned seriously. "Because you're wrong. You seem to have a talent for misinterpreting my actions, Nikki. But," he unscrewed the clasp of the necklace and put it back into the case, "if you don't want it—" He shrugged and held out his hand for the earrings.

Perversely, Nikki was annoyed that he wasn't trying to force her to accept the gift. Her refusal to fall in with his wishes had never stopped him before! She picked up a few magazines from the cocktail table, muttered good night, and climbed the stairs to Lauren's room.

After an unpromising start, the next day turned out to be a thorough success. At five o'clock in the morning, Mark and Jeffrey had burst into the master bedroom to wake their uncle. They were blasted by a furious lecture from the half-asleep Michael, who used

language that made even Nikki cringe. They had then
tiptoed down the hall into Lauren's room, and solemnly
informed their sister that Uncle Michael was in a
terrible mood and they had all better "shape up" or he
was never, ever going to come again.

Things improved considerably after that. Lauren
fixed pancakes for breakfast, indulging her brothers by
making them in the shapes of animals. Although
Michael's disposition was not precisely sunny, he didn't
scowl his way through the meal either. The twins were
irrepressibly active, but one stern look from their
mildly cranky uncle brought forth low "uh oh's" and
improved behavior whenever they threatened to become too unruly.

They all went hiking through the woods surrounding
the house in the morning, and Nikki watched jealously
as the other four scrambled up a tall spruce tree. It had
sprinkled during the night and everything smelled
earthy and fresh. Nikki inhaled the tangy pine scent
around her as if it were expensive perfume. She found
the nippy air intoxicating, and sadly eyed the cast on
her arm, which kept her from throwing out her hands
and whirling around joyfully as Laurie had done.

By the afternoon, Michael's good humor had been
restored. He suggested a softball game—three against
one—to his niece and nephews, who enthusiastically
agreed. He reined in his competitive instincts and let
them win. Afterward he closeted himself in the study
until dinner to make some phone calls; fortunately for
the worn-out Nikki, the children were content to watch
television.

Melanie had left some delicious stew that needed
only to be reheated. After an early dinner the exhausted Newman children were more than willing to drop
into bed. Michael said sardonically that at last he had

discovered the key to successful parenting. Keep them so busy that they're too tired to argue with you.

Nikki lasted only one hour longer than the children. She had exchanged few words with Michael during the day, and was unsure about whether they were on good terms with each other again. They were silently watching television when her eyes began to droop. "I'd carry you to your room, honey, but frankly I don't think I could make it up the stairs," Michael said wearily. Nikki was more than happy to bow to his suggestion that she go to sleep.

She woke up at three a. m. to the sound of someone rummaging around outside the house. Bravely deciding to investigate, she slipped her robe over her shoulders—the sleeve was too narrow to fit over her cast—and started down the stairs. She stopped halfway. Only reluctantly did she conclude that if a burglar were trying to break in, a slender young woman with one arm encased in plaster was not about to stop him. There was a clash of metal hitting concrete and her heart lurched.

She hurried back up the steps and through the hall to the master bedroom where Michael was sound asleep. "Michael!" she said in a soft, urgent voice. There was no response. She grabbed his bare shoulder and shook him.

He emitted a muffled protest and rolled onto his stomach. Nikki, abashed, realized that he was wearing absolutely nothing. She went over to the window and looked outside. There was a pounding sound, as if someone were hammering at a window frame, but she could make out no intruder.

A sleepy voice interrupted her examination of the back yard. "What's the matter?" Michael said thickly, and yawned.

"There's someone trying to get in. It woke me up," she hissed nervously.

Michael staggered out of bed and walked over to the window. Nikki supposed she should have blushed and looked away. Instead she stared at his muscled body, fascinated and aroused by his lean nakedness. He met her eyes, looked down absently, and with a grunt went to get a bathrobe. A sharp clatter erupted, and Nikki whispered, "There it is again!"

To her chagrin, Michael started to laugh at her. "Come on, city slicker," he mocked. "Let's go have a look at your vicious burglars."

He took her hand and led her down the stairs to the kitchen. The garbage cans by the rear entrance were visible through the large picture window. In the moonlight, Nikki could make out four black and white faces—two adults and two youngsters. A whole family of raccoons had expertly clawed the lids off the cans and was now feasting among the orange peels and steak bones.

"I feel like an idiot," Nikki moaned, as they watched the antics of the fat, stealthy robbers. "I really thought there were burglars—or worse."

"Lady," Michael replied drily, "if it's your virtue you're worried about, let me give you a piece of advice. Don't walk into gentlemen's bedrooms at three o'clock in the morning." He looked pointedly at her night-gown-clad body, which the robe was doing little to conceal.

Nikki was powerless to control the response of her senses to this aggressive examination. She only knew that she desperately wanted him to hold her. She took a step forward, halted. Then she glimpsed up from under her lashes and said huskily, "*Are* you a gentleman, Michael?"

He shot back, "I assume you're propositioning me, Nikki." Without giving her time to deny it, he slipped the robe from her shoulders, drew her close against the warmth of his body and began to tease her mouth with kitten-soft kisses. "What did I tell you would happen," he muttered against her lips, "if you provoked me again?" His tongue made a lazy tour of her mouth as his hands moved silkily over her stomach and breasts. Thoroughly aroused, Nikki stood on tiptoe and arched up against him, kissing him with abandon. Only when he swung her easily up off her feet did she have second thoughts. She hissed, "Put me down, Michael."

"Not on your life," he said evenly, heading toward the steps.

She struggled fitfully, sporadically, unable to make up her mind what she really wanted. Michael paused by the bed, still holding her high in his arms, to kiss her again. There was nothing lazy about his mouth now, only a brutal passion that excited and promised and frightened. When he tossed her roughly onto the bed and covered her body with his, all coherent thought ceased. She responded totally, her fingers gripping his hips, her breathing made shallow by the weight of his body on hers. There was nothing gentle or tentative in the way he pressed against her, his hands showing her how to move, his mouth ravaging her face, making her feel that she would scream if he didn't take her lips again.

Her gasping breathing must have penetrated his consciousness, because he abruptly rolled off her, firmly holding her body to bring her onto her side close against him. The suddenness of the action caught Nikki by surprise. Her cast-encased left hand, still gripping his hip, was unwittingly trapped under the full weight of his body.

She let out a yelp of pain. Michael sat up so quickly it was comical, and the two of them stared at each other in the moonlight.

"Damn," Michael said distinctly. "Damn, *damn,* DAMN!"

Nikki was stricken by irrational guilt. "I'm sorry, Michael, it just—"

"How the hell," he interrupted angrily, "am I supposed to make love to you— Get out of here, Nikki. And do me a favor, huh? Stay away from me from now on. Okay?"

It was the worst hurt Nikki had ever had to endure. She nodded dumbly, slunk back to Lauren's room, and cried herself to sleep.

The next morning, Nikki felt that she could gladly have slept forever. But with Lauren, Jeffrey and Mark badgering her to get up and oversee breakfast, she had no choice but to go downstairs. Laurie prepared bacon and eggs; Michael appeared just as the four of them were sitting down to eat. He muttered good morning to the kitchen wall and made himself a cup of instant coffee.

The two adults acted like wary boxers sparring for position. Neither confronted the other. Each of them developed an intense interest in the Sunday *Sun.* Eventually Michael suggested to his niece that she take charge of helping her brothers dress. "And see that it takes you at least fifteen minutes, Laurie," he added grumpily.

Laurie got the message. She shepherded the protesting twins out of the kitchen in a matter of seconds, barking orders at them like a determined collie. Once they were alone, Michael delivered himself of a disgusted sigh. It immediately put Nikki on the defensive.

"Let's get it over with," he said. "I know you'd

rather not have to see me today, but neither of us has any choice. So I'm sorry. I should have—hell, I don't know what I should have done."

Nikki felt compelled to make some comment. "I shouldn't have provoked you," she muttered into her hot chocolate.

"True."

"I—I guess you warned me," she added miserably.

"Also true."

"It could have been any woman and you would have reacted the same way."

"If you say so," Michael agreed blandly.

Nikki peeked up at him. "Wouldn't you?"

He shook his head. "Nikki, you have a lot to learn about men and even more to learn about me. But I'll forego the lectures until your wrist is healed. I'm not making the same mistake twice. Now. Are we on speaking terms again?"

Nikki ached to ask him why he had been so brutally angry with her last night but didn't dare. She nodded her agreement.

"Good," Michael said with a yawn. "Now where's the local paper? There must be something around here we can take the kids to today."

Lauren, Jeffrey and Mark were flinging harsh epithets at their implacable uncle. "Slavedriver!" "Straw boss!" "Meanie!" And worst of all: "You're just like daddy!"—were hurled at his unrelenting head. They had just returned from a local fair, where Uncle Michael had taken them on half a dozen rides, tramped around looking at animals, needlepoint, pottery and woodworking, and succumbed to their importunities for popcorn, cotton candy and peanuts. They declared unanimously that he was the best uncle in the world.

Nonetheless, to their way of thinking it was sheer

audacity for him to insist that they do their homework and help him clean up before their parents arrived home. When these chores were completed to his satisfaction, he reinstalled himself in their good graces by taking them out to the local hamburger place for dinner. "Over-done hamburgers may be the worst part of fatherhood," he told Nikki, wincing at the limp pickle slice resting atop his cheeseburger.

By the time Melanie and Brandt came home, three squeaky-clean, pajamaed offspring and a spotless house awaited them.

"Michael, you're disgustingly efficient," his older sister complained.

Brandt smiled at Michael and Nikki and told them warmly, "Thank you both for watching the brood this weekend. I'm sure it was a real treat for them. And some day I hope we can return the favor, Mike."

"And I'll help babysit your children," announced Lauren knowingly. "After you two are married, of course."

"Can't think of anyone I'd trust more," replied her uncle, and drawled at Nikki, "Isn't that right, darling?"

She wasn't about to rise to the bait. "If you say so, Michael," she agreed demurely.

As they drove back to the city, Nikki politely thanked Michael for taking her along for the weekend. She told him that she admired Melanie's ability to combine a family and career so smoothly.

"What about you?" Michael asked in a detached way. "Are you interested in marriage? Or kids?"

Nikki secretly thought that if she couldn't have Michael, she would never marry at all. "My career comes first. I suppose I'll get married some day—when I'm older."

"And if your husband wants children?" Michael

asked indifferently, apparently far more interested in the parkway than in her answer.

"Maybe when I'm in my thirties," Nikki lied, striving to mimic his lazy tone. "I guess I could get a full-time housekeeper and go back to work right away."

"I'll have to introduce you to some of my wealthy acquaintances, Miss Warren," Michael mocked gently. "I can see you've acquired an appreciation for luxury in the last—what is it?—seven weeks?"

"I get my cast off Friday." Nikki tried to sound casual. "And the sublease on my apartment expires on Wednesday. Will you help me move over the weekend?"

"I'll be in California. I have to leave Tuesday morning. I want you to wait until I get back—I want to make sure everything's okay first."

"Really, I can manage. I've—I've imposed on your hospitality long enough," Nikki insisted, grateful that the blackness of the night hid her ashen face. She had given Michael a clear opportunity to ask her to stay and he hadn't taken her up on it.

"You really want to leave, Nikki? Won't you miss—all the conveniences my money's been buying you?"

No, she thought. But I'll miss you. Horribly. Suddenly it seemed imperative not to postpone the inevitable. "Sure. But it's about time I started taking care of myself again." She licked her lips nervously. "Michael, I do want to thank you again. I really appreciate the way you—"

"For God's sake," he barked, "don't start riding that horse again!" He reached over to click on the radio, and the strains of discordant modern music filled the car.

Chapter 10

Nikki lay in bed, and cocked her wrist absent-mindedly
back and forth, ignoring the soreness and stiffness. It
was no use. She simply couldn't sleep. Her cast had
been removed, her apartment was vacant. Although
nothing more had been said about her moving, she
knew there was no longer any reason for her to remain
at Michael Cragun's apartment.

Her thoughts refused to leave the man with whom
she was so much in love. During the past month, even
though Michael was out of town a great deal, she had
spent many hours in his company. She had observed
him in myriad moods, and on one too-brief occasion,
she told herself ruefully, he had been a passionate
lover. She was tormented by the memory of those
moments when she had been lost in his arms.

She was assailed by an overwhelming desire to be
somehow closer to him. This would be the last night she
would spend in his home. She would leave tomorrow
while he was still in California.

She left her own room and crossed the hall into his,
then snuggled down into his bed. It was not enough.
She found his bulky terry cloth robe hanging on a hook
on the back of the bathroom door. She slipped it on
over her flimsy apricot nightgown and cuddled it close
to her face. The scent reminded her of Michael. Then

she nestled back into his bed, the large robe enveloping her like a cocoon.

Michael had left for the West Coast before she had awakened on Tuesday, so there had been no good-byes between them. Nikki thought it was probably for the best; she could not endure the humiliation of bursting into tears in front of him. She had hoped fruitlessly for some further word from him—not even necessarily of affection, just that he would miss her, or that he wanted her to stay a little longer. None was forthcoming. In all fairness, she admitted that her own attitude—withdrawn and defensive—could hardly have encouraged him. He would probably be glad to get rid of her.

She woke up when she felt something touch her neck. Groggy with sleep, it took her a moment to realize that there was someone in bed with her, and then she opened her mouth to scream. A hand was clapped firmly over her parted lips and a familiar voice drawled into her ear, "And who's been sleeping in *my* bed? Not Goldilocks—wrong color hair. Are you the virginal Snow White by any chance?"

Sunrise was half an hour away. Now that Nikki's eyes were accustomed to the dark, enough light filtered into the room through the heavy brown drapes for Nikki to see Michael lying in bed close to her, his chest bare. She felt that she had to attempt some sort of explanation; she could scarcely say, "I'm so in love with you I crawled into your bed to pretend you were with me."

"I couldn't sleep," she mumbled. "Sometimes—a different bed helps."

"Umm. I see. Did the robe help also?" he queried, not believing her.

"I was cold," she lied, now anything but cold as the feel of his hip against hers registered on her senses.

"I'll have to see what I can do about that, won't I?" he laughed softly. He slid gently on top of her, careful to support his weight with his elbow, and she realized with a certain relief that he was wearing pajama bottoms. She didn't struggle or stiffen. She wanted him too much.

Michael opened up the bulky robe she was wearing and began to gently caress her body. He made no move to kiss her, but regarded her face intently as his hand moved across her shoulders, brushing away the spaghetti straps of her nightgown, then pulling it down to expose her body. He traced the bones around her neck with a teasing touch, then moved his fingers lower to stroke her already taut nipples.

For a little while Nikki stared tremblingly back into his eyes as he watched her, but when the sensations he was arousing became too intense, she gave a little moan, closed her eyes, and reached up her own hand to run it over his shoulders and back.

She felt him lift up the other arm, the newly healed left one, and tenderly kiss her wrist. "Is it all right? I don't want to hurt you again," he said hoarsely.

"It's fine," she whispered back. "Don't stop. I won't break."

Nonetheless, when he finally bent his head to kiss her, he was so guarded and wary that after a few minutes Nikki felt frustrated enough to bite him. She grabbed his hair to bring him closer, and began to kiss him back wantonly. Michael soon forgot about her recent injury, and, breathing raggedly, he took control in an aggressive, demanding manner. He pulled Nikki on top of him and tangled one hand in her hair to position her head where he wanted it. Arms trapped underneath him, she could only respond to the directions of his other hand, which was splayed across the small of her back.

Nikki had never imagined that she could crave complete release this way. She arched her body against his so that the two of them fit together like lock and key, and obeyed his unspoken commands without thought of rebellion. When he reached up to slip her long nightgown up over her hips, she helped him impatiently.

And then he stopped. Michael pushed her away from him and stalked over to the window, parting the drapes and staring down across the park.

Tears of frustration filled her eyes. "Michael, what's wrong? What did I do? Just teach me—please," she gulped. "I'll do—whatever you tell me."

His answer was snapped out. "No! You didn't do anything *wrong*. Just get out of here, Nikki. While I can still let you go. Go to your own bedroom. Now!"

"But—but, Michael—" she pleaded brokenly.

"*Now!*" he shouted.

Nikki ran from the room, fled down the hall into her own, and slammed the door behind her. She was stunned by his rejection and completely baffled by his behavior. Why had he stopped so abruptly? Surely her own willingness had been apparent. There was no plaster cast to dampen his passion. And he had wanted her. She might be inexperienced, but she knew enough to be able to tell *that!*

But as her tears dried and her body cooled, she told herself defeatedly that he had done her a favor. It seemed an age ago that he had told her that although he was attracted to her and cared for her, he didn't love her. She could hardly blame him for responding to the presence of a willing young woman in his bed. But fortunately for her sake, he had had the decency and common sense to stop.

She knew she could never face him—her actions must have made her feelings perfectly plain. She pulled down a suitcase from the shelf of her closet and threw some clothes into it. Henry could bring the rest of her things later. Then she quickly pulled on jeans, a turtleneck and jacket, and started down the stairs.

She was a few feet from the front door when Michael shouted angrily down at her, "Just where do you think you're going? I'm not letting you run away the minute my back is turned!"

She made no reply, and was turning the handle of the door when he came hurtling down the steps three at a time. He grabbed the suitcase from her, slammed the door closed and spun her around.

Hands biting into her shoulders, he told her furiously, "You've been driving me nuts since almost the first moment I met you, you know that? You tease me until I'm crazy with wanting you but you won't listen to anything I tell you about my feelings." He gave her a fierce little shake and then challenged, "I may be wrong but I think I do know something about women. I want you to stand there and tell me you don't love me. And maybe I'll even believe you!"

Her whole body sagged. "Is it going to give you that much satisfaction to hear me say it?" she asked resignedly. "Finc. I love you. You've had your fun. Now let me go."

He gave a deep sigh, and looked almost relieved. "For such a smart girl," he said tenderly, "you certainly can be stupid. Don't you know—after just now—how much I love you?"

Seeking reassurance, Nikki whispered, "No. Last weekend, you couldn't wait to be rid of me. And just now, you threw me out—and I wanted to stay."

"Nikki, for God's sake, do you think I didn't want you to? Both times? But last week I was afraid I'd break your bloody wrist again. Another six weeks of this and I'd be certifiably insane. And tonight—I remembered our talk about children—it occurred to me

I could get you pregnant. Hell, I needed time to cool off. You *do* know where babies come from, don't you?" he mocked.

His words were so ironic in view of the bizarre way they had met that Nikki couldn't help giggling. Once she started, she couldn't stop, and it was infectious. Michael was still laughing as he said, "Okay, okay. It's poetic justice, I admit it."

Then he took her arm and led her into his study, gently pushing her down onto the brown leather couch. He opened his wall safe and removed a small box.

"Open it," he ordered.

Inside was a ring: a brilliant square-cut diamond surrounded by smaller round emeralds.

"Nikki," Michael said with uncharacteristic nervousness, "let's fly to Las Vegas later. We can go to Hawaii afterward if that's okay with you."

"Aren't you forgetting something?" she asked impishly.

"Huh?" He was genuinely puzzled. Then a light dawned. "Oh! You mean the necklace and earrings. Wait, I'll get them." He headed back to the safe.

"No, you arrogant beast," she said, exasperated. "You might at least *ask* me to marry you!"

He shook his head. "I thought I had." Returning to the couch, he took her head between his hands. "Marry me, Nikki. I love you—very, very much." He ran his lips along her neck, then kissed her mouth gently. Then he leaned back and grinned at her. "Satisfied?"

"Umm. Although Lauren won't be. She expects to be a bridesmaid." She slipped the ring on.

He was suddenly solemn. "I've put you through a miserable time, Nikki. Until tonight, I was afraid that the attraction on your part was purely physical. But when I saw you in my bed, wearing my robe—I didn't think you would do something so sentimental unless you cared. And I watched the expression in your eyes while I was making love to you. And I knew."

"Why did you stop?" Nikki asked for the second time.

"I told you. You were so negative about kids. When you're ready for them—if you ever want them—you tell me, honey. I'd like children, but I don't want to push you."

"Then it's the first time you've shown such scruples!" Nikki mocked. "It just so happens that I love kids— especially the ones we're going to have together. But if I didn't, you'd probably bulldoze me into motherhood anyway." She snuggled contentedly close to him. "I don't understand why you never said anything—after the time in the hospital. You've treated me so much like a little sister these past weeks—except when I practically propositioned you—that I thought that's how you regarded me."

"Little sister, indeed!" he snorted. "I assure you, Miss Warren, you don't need to provoke me to make me want to haul you off to bed." He paused to prove it in a very acceptable manner, before continuing softly, "But every time I tried to show you or tell you how I felt, you refused to believe me. I figured you needed time to get to know me. I wanted to take you places, to—court you, if that isn't too old-fashioned. Besides, with that left arm of yours all plastered up, I decided it would be better to wait. You made it almost impossible for me, although I admit I kind of enjoyed the game you were playing. But I *don't* enjoy restrained, polite love-making and it had me climbing the walls to have to stop after I nearly rebroke your arm."

Nikki was gratified by that last statement. But his recital was far from complete enough for her. She wanted a chronicle of his emotions, right from the start.

She began, "The night of your dinner party, I—overheard you talking to Elianna. She said you'd had a fight over having children."

"You listened? You nosy little wretch." He grinned at her. "Do it again and I'll tie you to the bed and violate your pure little body!"

"Promise, Michael?" Nikki cooed.

"Absolutely," he winked.

"Elianna?"

"Cold and selfish, but with—uh—compensations. Incapable of fidelity though. I did know about the conductor, by the way. I suppose a rich husband appealed to her. When she started nagging about marriage, I tossed her a line about how important kids are to me. I knew she couldn't stand them. She told me if a baby machine was what I wanted, I should advertise for one. It was a dare, and frankly the whole idea appealed to my jaded sense of humor. I steam-rollered Charlie into handling it for me."

"You told her I was boring—" Nikki pouted.

"What?"

"You said that I was just a child. And then you said that a certain type of female bored you senseless, and—I didn't stay to listen to the rest."

"Is that why you were so rotten to me all week?" he asked, incredulous. Nikki nodded, her eyes sparkling with amusement. "You should have stuck around. I told Elianna that jealous, nagging women were stultifying, and gave her the choice of being thrown out—after all, I hadn't invited her—or behaving herself. I also told her I wouldn't be seeing her again—except on a stage."

"But you'd just spent the weekend with her," Nikki said sourly.

"Jealous?"

"I shouldn't admit it, but the truth is I was disgustingly jealous. And I realized I was in love with you. Do you mind? Because you just said you can't stand jealous women," Nikki answered with a concerned frown on her face.

"In your case, I'm going to have to make an exception. I felt like killing your boss when I found the two of you together." He kissed her palm. "After my unfortunate encounter with your knee, I deliberately set out to gain your trust. That whole line about taking care of you—being your big brother—it was a bunch of bull. I didn't know what I wanted you to be, but it sure wasn't a little sister!"

"I actually believed you! You're an evil man, Michael," Nikki accused.

"Don't make faces at me, honey. I was properly punished. I wouldn't admit to myself that I was falling in love with you. That weekend with Elianna—I guess I wanted you to be jealous. But you had your revenge."

Nikki studied his expression, trying to figure out what on earth he meant. To her astonishment, he was actually blushing. "Well, don't stop now," she said, fascinated.

"After you left that snake of yours in my bed—after your Uncle Michael number in my bedroom—I intended to sleep with Ellie. I think I wanted to prove to myself that you didn't matter. But when I closed my eyes, I saw your face, not hers, and I couldn't—" he halted abruptly.

"Couldn't what?" she asked with a smirk.

"Just *couldn't,* you little cat. That's never happened before. I ended up taking her home twenty minutes later. I'm only thirty-two. I mean—I know I'm not nineteen anymore, even though for the past month I've felt like it. But still—"

"It's what you deserve," Nikki said smugly. "And if

you want the truth, now that I know you've been frustrated for the last month, I'm absolutely delighted with my revenge!"

Michael's answer was to pin her down and begin to kiss and caress her until she was moaning and clutching at him. Then he roughly pushed her away and growled, "Don't play games with me, lady! Now you know how I've felt every bloody night for the last six weeks. Why do you think I've been traveling so much? When I've been home, knowing you were across the hall, taunting me with that seductive routine of yours—I bought the ring at the same time as the necklace and earrings. I was going to give it to you when I got back from California Monday night. That's why I told you not to leave. But all day Saturday I thought of you, here, the cast off—and I didn't want to wait. I took the company plane back to New York and spent the whole trip trying to figure out how I was going to convince you to marry me. I know I could have seduced you any time, Nikki," he told her bluntly, "but it wasn't what I wanted. Even in the beginning I knew I could get you into bed, but I had this feeling you would hate me afterward and I couldn't live with that."

"But in the car, before the accident—"

"Damn it, you'd pushed me to the point where I didn't care any more. That last minute attack of conscience—it wouldn't have mattered. I wanted you so much that nothing would have kept me out of your bedroom that night. The hurt look on your face—I already half-knew I was in love with you. And then that truck hit us—" He slowly shook his head. "Six wasted weeks."

Nikki had one last question. "Michael, why did you—bother—at first, I mean? You saw I didn't want to go through with it."

"I was intrigued by you before we even met. When Charlie told me about your reply to the ad, I sent a whole team of people out to investigate you. You saw how complete the dossier was. And then once I saw you, I didn't want to let you go. I suppose it was another game. It isn't very pleasant to admit this, but I enjoyed—manipulating you. I was so sick of spoiled, rich, selfish women, and you were young and vulnerable, and beautiful. And best of all, you fought me—you were a challenge." His face mirrored the tone of self-reproach. "I got a kick out of forcing you to do what I wanted. Cat and mouse. I knew you'd walk out after we got back from Florida. I relished—timing the pressure, forcing you to come back. I really don't know if I would have carried out all the threats. I don't like to think I would have gone that far."

He began to absent-mindedly play with her hair, saying softly, "I told you after the accident that I'd set you up—that I felt guilty. But you wouldn't let me explain that when you lay unconscious in that car, waiting for the ambulance, and your blood was all over the place—oh God, I sat there, holding you, wishing it was me, realizing how much I loved you. But when I brought up how I felt, you seemed determined not to believe me. I think I must have loved you almost from the beginning—from the time we spent in Florida. It made me determined to hold onto you. But after the accident, I figured you had to hate my guts."

His face wore an agonized expression, as if he were remembering the blood, and her pain. "Nikki, I—"

Abruptly Nikki lashed out at him. "Shut up!"

He stared at her, his face pale. "Has this all been some kind of revenge? I don't understand—"

"Mr. Cragun," she interrupted coldly, "shut up and do as you're told. Get upstairs—*Now!*"

A slow smile replaced the anxious look on his face. "Yes, ma'am," he said obediently.

"You know," she said teasingly, "I rather fancy the idea of having your child. Right away, Michael."

He kissed her nose, then picked her up and carried her up the stairs.